1992
YEARBOOK of
ASTRONOMY

1992
YEARBOOK of
ASTRONOMY.

edited by
Patrick Moore

Foreword by The Astronomer Royal

W. W. Norton & Company
NEW YORK   LONDON

First American Edition 1991

First Published in Great Britain 1991

Copyright © 1991 by Sidgwick and Jackson Ltd

All Rights Reserved.

ISBN 0-393-03078-4

W. W. Norton & Company, Inc.,
500 Fifth Avenue, New York, NY 10110
W. W. Norton & Company, Ltd,
10 Coptic Street, London WC1A 1PU

1 2 3 4 5 6 7 8 9 0

Photoset by Rowland Phototypesetting Ltd
Bury St Edmunds, Suffolk
Printed and bound in Great Britain by
Mackays of Chatham PLC, Chatham, Kent

Contents

Foreword by
The Astronomer Royal

It is a pleasure to write this short Foreword to the thirtieth *Yearbook of Astronomy*. The popularity of this series is bound up not only with the undoubted attractiveness of the subject, but also, and perhaps more particularly, with the charisma and skill of the subject's foremost proponent: Patrick Moore. We salute him as we recognize this important milestone in astronomy-publishing.

Without doubt, the *Yearbooks* have satisfied an important need. We have had an impressive collection of information for observers: the Star Charts and details of Astronomical Phenomena, the 'Monthly Notes'; data of such quality that even Astronomers Royal have been known to derive their information for public utterances therefrom. But that is not all, we have been treated to authoritative articles on a wide variety of exciting and topical subjects, from radio waves to gamma rays and from descriptions of planetary atmospheres to the latest in cosmological speculations. The 'other astronomies' have not been neglected: our knowledge about the cosmos from studies of meteorites and cosmic-ray particles has also come alive in these pages.

May the lifetime of the *Yearbook of Astronomy* series be truly astronomical.

ARNOLD WOLFENDALE, FRS
Astronomer Royal

Editor's Note

For the thirtieth edition of the *Yearbook* we are honoured with a Foreword by the Astronomer Royal, Professor Arnold Wolfendale, so I will be brief. The 1992 edition follows the traditional pattern, with many of our regular contributors and also some new ones; as usual, we have done our best to give a wide range, both of subject and of technical level. Also as usual, Gordon Taylor has provided all the data for the monthly notes.

PATRICK MOORE
Selsey, April 1991

Preface

New readers will find that all the information in this *Yearbook* is
given in diagrammatic or descriptive form; the positions of the
planets may easily be found on the specially designed star charts,
while the monthly notes describe the movements of the planets and
give details of other astronomical phenomena visible in both the
northern and southern hemispheres. Two sets of the star charts are
provided. The **Northern Charts** (pp. 14 to 39) are designed for use in
latitude 52 degrees north, but may be used without alteration
throughout the British Isles, and (except in the case of eclipses and
occultations) in other countries of similar north latitude. The
Southern Charts (pp. 40 to 65) are drawn for latitude 35 degrees
south, and are suitable for use in South Africa, Australia and
New Zealand, and other stations in approximately the same south
latitude. The reader who needs more detailed information will
find *Norton's Star Atlas* (Longman) an invaluable guide, while
more precise positions of the planets and their satellites, together
with predictions of occultations, meteor showers, and periodic
comets may be found in the *Handbook* of the British Astronomical
Association. The British monthly periodical, with current news,
articles, and monthly notes is *Astronomy Now*. Readers will also
find details of forthcoming events given in the American *Sky and
Telescope*. This monthly publication also produces a special occultation supplement giving predictions for the United States and
Canada.

Important Note
The times given on the star charts and in the Monthly Notes are
generally given as local times, using the 24-hour clock, the day
beginning at midnight. All the dates, and the times of a few events
(e.g. eclipses), are given in Greenwich Mean Time (G.M.T.),
which is related to local time by the formula

Local Mean Time = G.M.T. − west longitude

In practice, small differences of longitudes are ignored, and the
observer will use local clock time, which will be the appropriate

Standard (or Zone) Time. As the formula indicates, places in west longitude will have a Standard Time slow on G.M.T., while places in east longitude will have a Standard Time fast on G.M.T. As examples we have:

Standard Time in

New Zealand	G.M.T.	+	12 hours
Victoria; N.S.W.	G.M.T.	+	10 hours
Western Australia	G.M.T.	+	8 hours
South Africa	G.M.T.	+	2 hours
British Isles	G.M.T.		
Eastern S.T.	G.M.T.	−	5 hours
Central S.T.	G.M.T.	−	6 hours, etc.

If Summer Time is in use, the clocks will have to have been advanced by one hour, and this hour must be subtracted from the clock time to give Standard Time.

In Great Britain and N. Ireland, Summer Time will be in force in 1992 from March $29^{d}01^{h}$ until October $25^{d}01^{h}$ G.M.T.

Notes on the Star Charts

The stars, together with the Sun, Moon and planets seem to be set on the surface of the celestial sphere, which appears to rotate about the Earth from east to west. Since it is impossible to represent a curved surface accurately on a plane, any kind of star map is bound to contain some form of distortion. But it is well known that the eye can endure some kinds of distortion better than others, and it is particularly true that the eye is most sensitive to deviations from the vertical and horizontal. For this reason the star charts given in this volume have been designed to give a true representation of vertical and horizontal lines, whatever may be the resulting distortion in the shape of a constellation figure. It will be found that the amount of distortion is, in general, quite small, and is only obvious in the case of large constellations such as Leo and Pegasus, when these appear at the top of the charts, and so are drawn out sideways.

The charts show all stars down to the fourth magnitude, together with a number of fainter stars which are necessary to define the shape of a constellation. There is no standard system for representing the outlines of the constellations, and triangles and other simple figures have been used to give outlines which are easy to follow with the naked eye. The names of the constellations are given, together with the proper names of the brighter stars. The apparent magnitudes of the stars are indicated roughly by using four different sizes of dots, the larger dots representing the brighter stars.

The two sets of star charts are similar in design. At each opening there is a group of four charts which give a complete coverage of the sky up to an altitude of 62½ degrees; there are twelve such groups to cover the entire year. In the **Northern Charts** (for 52 degrees north) the upper two charts show the southern sky, south being at the centre and east on the left. The coverage is from 10 degrees north of east (top left) to 10 degrees north of west (top right). The two lower charts show the northern sky from 10 degrees south of west (lower left) to 10 degrees south of east (lower right). There is thus an overlap east and west.

Conversely, in the **Southern Charts** (for 35 degrees south) the upper two charts show the northern sky, with north at the centre and east on the right. The two lower charts show the southern sky, with south at the centre and east on the left. The coverage and overlap is the same on both sets of charts.

Because the sidereal day is shorter than the solar day, the stars appear to rise and set about four minutes earlier each day, and this amounts to two hours in a month. Hence the twelve groups of charts in each set are sufficient to give the appearance of the sky throughout the day at intervals of two hours, or at the same time of night at monthly intervals throughout the year. The actual range of dates and times when the stars on the charts are visible is indicated at the top of each page. Each group is numbered in bold type, and the number to be used for any given month and time is summarized in the following table:

Local Time	18h	20h	22h	0h	2h	4h	6h
January	11	12	1	2	3	4	5
February	12	1	2	3	4	5	6
March	1	2	3	4	5	6	7
April	2	3	4	5	6	7	8
May	3	4	5	6	7	8	9
June	4	5	6	7	8	9	10
July	5	6	7	8	9	10	11
August	6	7	8	9	10	11	12
September	7	8	9	10	11	12	1
October	8	9	10	11	12	1	2
November	9	10	11	12	1	2	3
December	10	11	12	1	2	3	4

The charts are drawn to scale, the horizontal measurements, marked at every 10 degrees, giving the azimuths (or true bearings) measured from the north round through east (90 degrees), south (180 degrees), and west (270 degrees). The vertical measurements, similarly marked, give the altitudes of the stars up to 62½ degrees. Estimates of altitude and azimuth made from these charts will necessarily be mere approximations, since no observer will be exactly at the adopted latitude, or at the stated time, but they will serve for the identification of stars and planets.

The ecliptic is drawn as a broken line on which longitude is marked at every 10 degrees; the positions of the planets are then easily found by reference to the table on page 71. It will be noticed that on the Southern Charts the **ecliptic** may reach an altitude in excess of 62½ degrees on star charts 5 to 9. The continuations of the broken line will be found on the charts of overhead stars.

There is a curious illusion that stars at an altitude of 60 degrees or more are actually overhead, and the beginner may often feel that he is leaning over backwards in trying to see them. These overhead stars are given separately on the pages immediately following the main star charts. The entire year is covered at one opening, each of the four maps showing the overhead stars at times which correspond to those of three of the main star charts. The position of the zenith is indicated by a cross, and this cross marks the centre of a circle which is 35 degrees from the zenith; there is thus a small overlap with the main charts.

The broken line leading from the north (on the Northern Charts) or from the south (on the Southern Charts) is numbered to indicate the corresponding main chart. Thus on page 38 the N-S line numbered 6 is to be regarded as an extension of the centre (south) line of chart 6 on pages 24 and 25, and at the top of these pages are printed the dates and times which are appropriate. Similarly, on page 65, the S-N line numbered 10 connects with the north line of the upper charts on pages 58 and 59.

The overhead stars are plotted as maps on a conical projection, and the scale is rather smaller than that of the main charts.

1L

October 6 at 5ʰ | October 21 at 4ʰ
November 6 at 3ʰ | November 21 at 2ʰ
December 6 at 1ʰ | December 21 at midnight
January 6 at 23ʰ | January 21 at 22ʰ
February 6 at 21ʰ | February 21 at 20ʰ

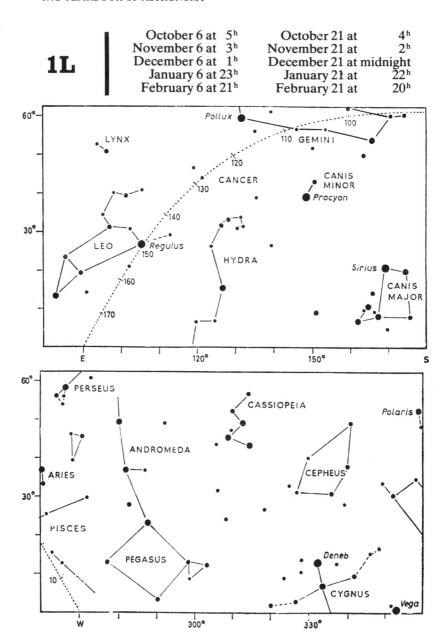

October 6 at 5^h	October 21 at	4^h
November 6 at 3^h	November 21 at	2^h
December 6 at 1^h	December 21 at midnight	
January 6 at 23^h	January 21 at	22^h
February 6 at 21^h	February 21 at	20^h

1R

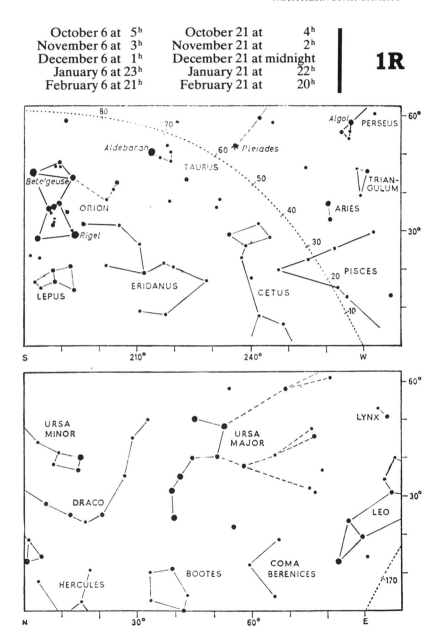

2L

November 6 at 5ʰ	November 21 at 4ʰ
December 6 at 3ʰ	December 21 at 2ʰ
January 6 at 1ʰ	January 21 at midnight
February 6 at 23ʰ	February 21 at 22ʰ
March 6 at 21ʰ	March 21 at 20ʰ

November 6 at 5ʰ   November 21 at 4ʰ
December 6 at 3ʰ   December 21 at 2ʰ
January 6 at 1ʰ   January 21 at midnight
February 6 at 23ʰ   February 21 at 22ʰ
March 6 at 21ʰ   March 21 at 20ʰ

2R

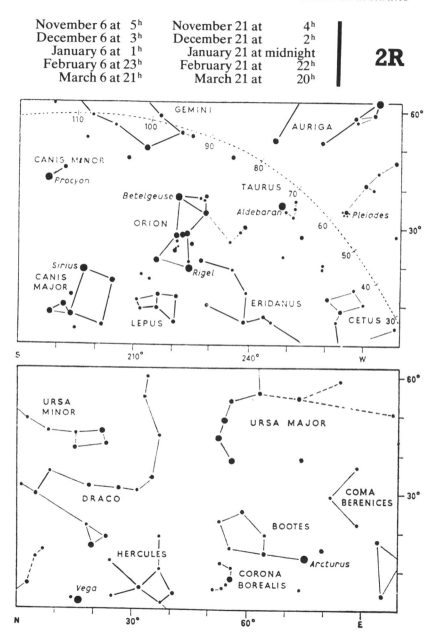

3L

December 6 at 5^h	December 21 at 4^h
January 6 at 3^h	January 21 at 2^h
February 6 at 1^h	February 21 at midnight
March 6 at 23^h	March 21 at 22^h
April 6 at 21^h	April 21 at 20^h

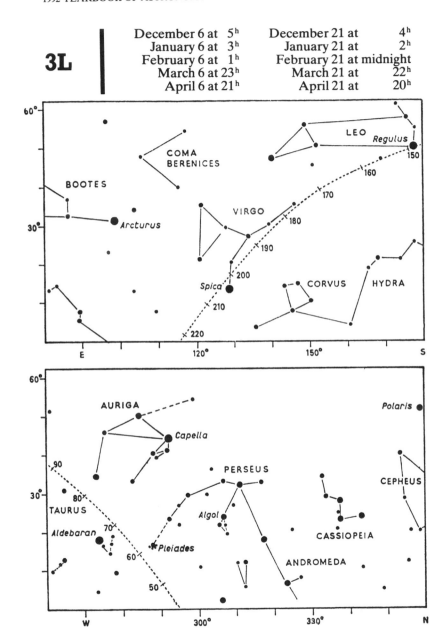

December 6 at 5^h	December 21 at 4^h	
January 6 at 3^h	January 21 at 2^h	
February 6 at 1^h	February 21 at midnight	**3R**
March 6 at 23^h	March 21 at 22^h	
April 6 at 21^h	April 21 at 20^h	

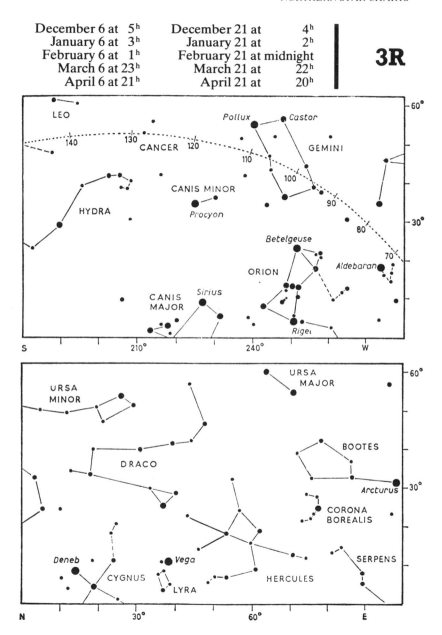

4L

January 6 at 5^h	January 21 at 4^h
February 6 at 3^h	February 21 at 2^h
March 6 at 1^h	March 21 at midnight
April 6 at 23^h	April 21 at 22^h
May 6 at 21^h	May 21 at 20^h

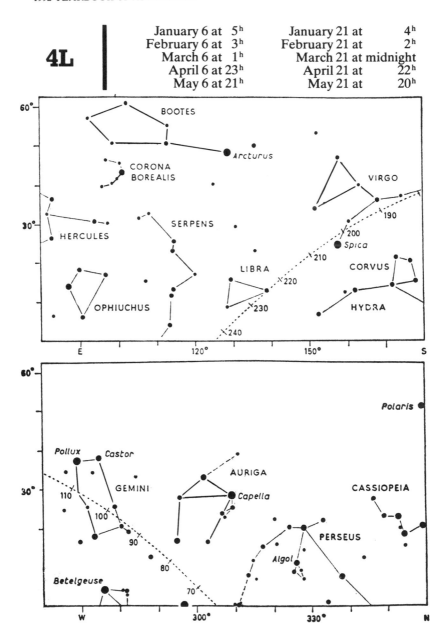

January 6 at 5ʰ   January 21 at   4ʰ
February 6 at 3ʰ   February 21 at   2ʰ
March 6 at 1ʰ   March 21 at midnight
April 6 at 23ʰ   April 21 at   22ʰ   **4R**
May 6 at 21ʰ   May 21 at   20ʰ

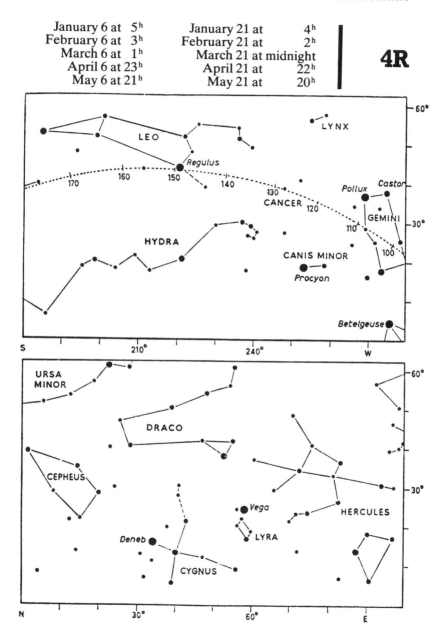

5L

January 6 at 7ʰ   January 21 at   6ʰ
February 6 at 5ʰ   February 21 at   4ʰ
March 6 at 3ʰ   March 21 at   2ʰ
April 6 at 1ʰ   April 21 at midnight
May 6 at 23ʰ   May 21 at   22ʰ

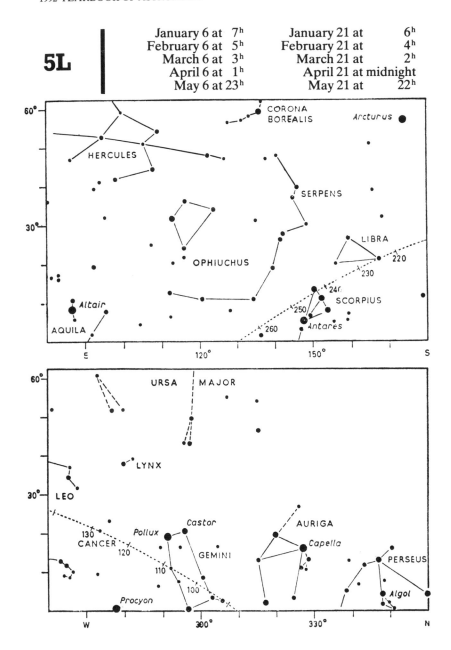

January 6 at 7ʰ	January 21 at 6ʰ
February 6 at 5ʰ	February 21 at 4ʰ
March 6 at 3ʰ	March 21 at 2ʰ
April 6 at 1ʰ	April 21 at midnight
May 6 at 23ʰ	May 21 at 22ʰ

5R

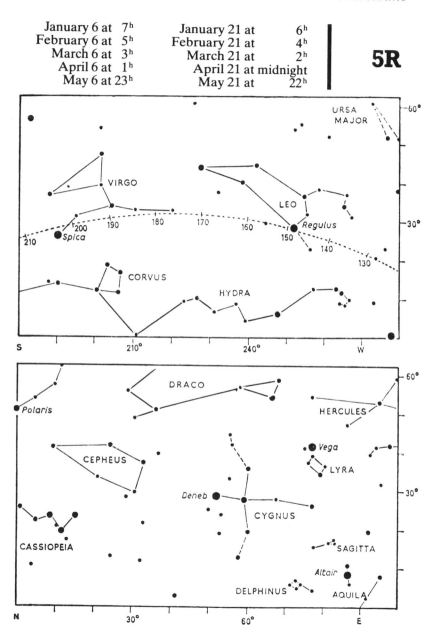

23

6L

March 6 at 5ʰ   March 21 at   4ʰ
April 6 at 3ʰ   April 21 at   2ʰ
May 6 at 1ʰ   May 21 at midnight
June 6 at 23ʰ   June 21 at   22ʰ
July 6 at 21ʰ   July 21 at   20ʰ

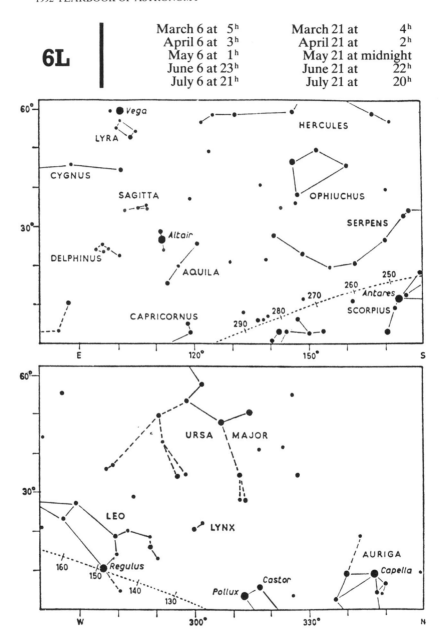

March 6 at 5^h March 21 at   4^h
April 6 at 3^h April 21 at   2^h
May 6 at 1^h May 21 at midnight   **6R**
June 6 at 23^h June 21 at   22^h
July 6 at 21^h July 21 at   20^h

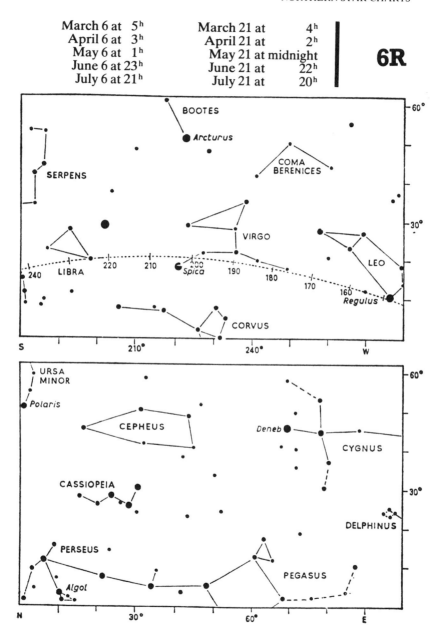

7L

May 6 at 3ʰ
June 6 at 1ʰ
July 6 at 23ʰ
August 6 at 21ʰ
September 6 at 19ʰ

May 21 at   2ʰ
June 21 at midnight
July 21 at   22ʰ
August 21 at   20ʰ
September 21 at   18ʰ

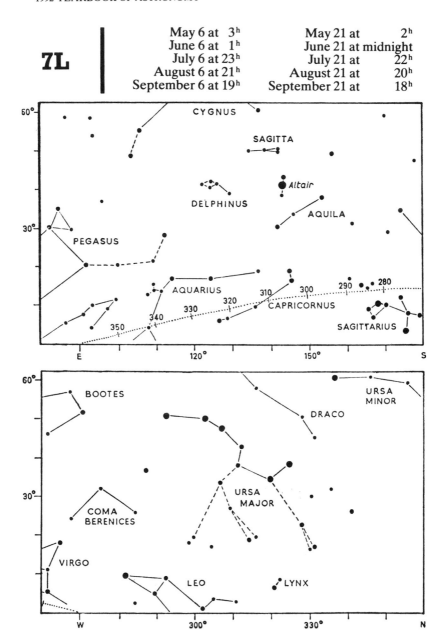

May 6 at 3[h]	May 21 at 2[h]	
June 6 at 1[h]	June 21 at midnight	
July 6 at 23[h]	July 21 at 22[h]	**7R**
August 6 at 21[h]	August 21 at 20[h]	
September 6 at 19[h]	September 21 at 18[h]	

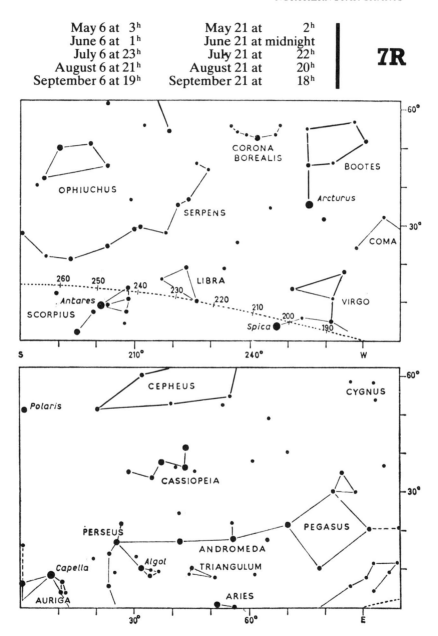

8L

July 6 at 1ʰ	July 21 at midnight
August 6 at 23ʰ	August 21 at 22ʰ
September 6 at 21ʰ	September 21 at 20ʰ
October 6 at 19ʰ	October 21 at 18ʰ
November 6 at 17ʰ	November 21 at 16ʰ

July 6 at 1ʰ	July 21 at midnight
August 6 at 23ʰ	August 21 at 22ʰ
September 6 at 21ʰ	September 21 at 20ʰ
October 6 at 19ʰ	October 21 at 18ʰ
November 6 at 17ʰ	November 21 at 16ʰ

8R

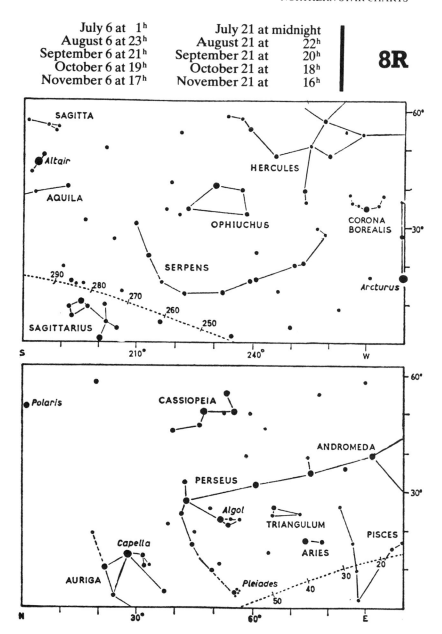

9L

August 6 at 1ʰ	August 21 at midnight
September 6 at 23ʰ	September 21 at 22ʰ
October 6 at 21ʰ	October 21 at 20ʰ
November 6 at 19ʰ	November 21 at 18ʰ
December 6 at 17ʰ	December 21 at 16ʰ

August 6 at   1ʰ   August 21 at midnight
September 6 at 23ʰ   September 21 at   22ʰ
October 6 at 21ʰ   October 21 at   20ʰ   **9R**
November 6 at 19ʰ   November 21 at   18ʰ
December 6 at 17ʰ   December 21 at   16ʰ

10L

August 6 at 3^h	August 21 at 2^h
September 6 at 1^h	September 21 at midnight
October 6 at 23^h	October 21 at 22^h
November 6 at 21^h	November 21 at 20^h
December 6 at 19^h	December 21 at 18^h

August 6 at 3^h	August 21 at 2^h	
September 6 at 1^h	September 21 at midnight	**10R**
October 6 at 23^h	October 21 at 22^h	
November 6 at 21^h	November 21 at 20^h	
December 6 at 19^h	December 21 at 18^h	

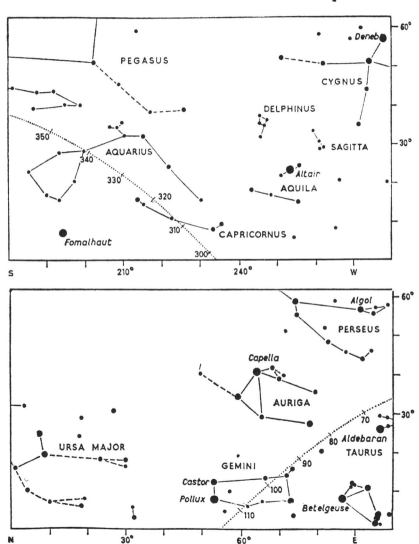

11L

September 6 at 3ʰ	September 21 at 2ʰ
October 6 at 1ʰ	October 21 at midnight
November 6 at 23ʰ	November 21 at 22ʰ
December 6 at 21ʰ	December 21 at 20ʰ
January 6 at 19ʰ	January 21 at 18ʰ

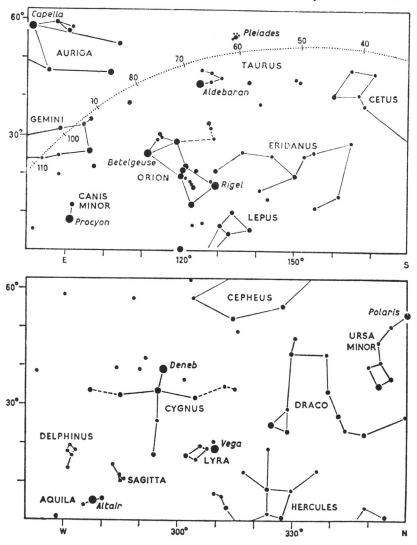

September 6 at 3^h September 21 at   2^h
October 6 at 1^h October 21 at midnight
November 6 at 23^h November 21 at   22^h **11R**
December 6 at 21^h December 21 at   20^h
January 6 at 19^h January 21 at   18^h

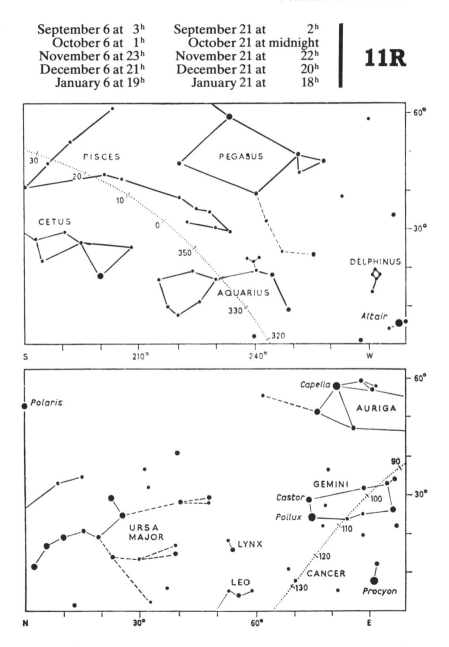

12L

October 6 at 3ʰ	October 21 at 2ʰ
November 6 at 1ʰ	November 21 at midnight
December 6 at 23ʰ	December 21 at 22ʰ
January 6 at 21ʰ	January 21 at 20ʰ
February 6 at 19ʰ	February 21 at 18ʰ

October 6 at 3^h October 21 at   2^h
November 6 at 1^h November 21 at midnight
December 6 at 23^h December 21 at   22^h
January 6 at 21^h January 21 at   20^h
February 6 at 19^h February 21 at   18^h

12R

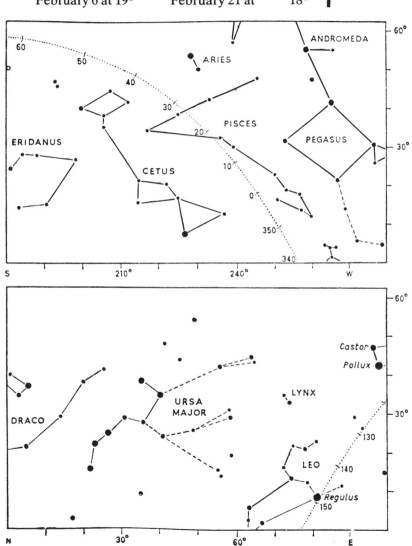

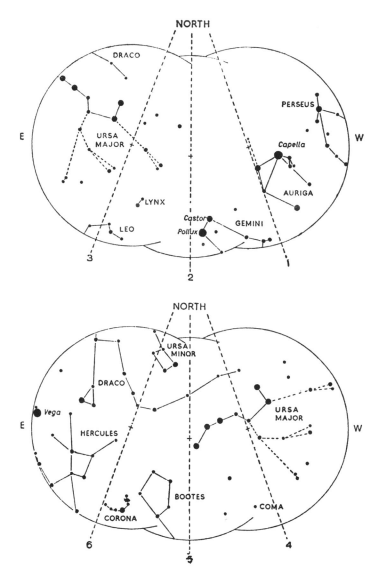

Northern Hemisphere Overhead Stars

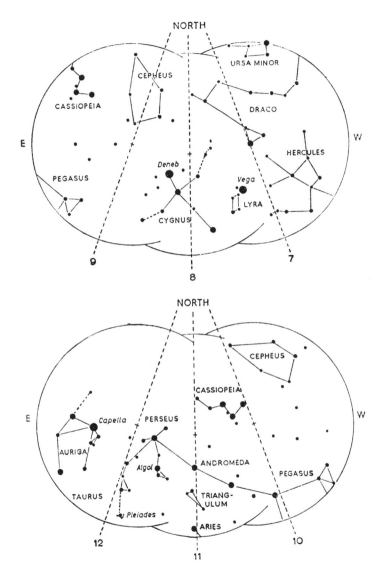

Northern Hemisphere Overhead Stars

1L

October 6 at 5ʰ	October 21 at 4ʰ
November 6 at 3ʰ	November 21 at 2ʰ
December 6 at 1ʰ	December 21 at midnight
January 6 at 23ʰ	January 21 at 22ʰ
February 6 at 21ʰ	February 21 at 20ʰ

October 6 at   5ʰ   October 21 at   4ʰ
November 6 at   3ʰ   November 21 at   2ʰ
December 6 at   1ʰ   December 21 at midnight   **1R**
January 6 at 23ʰ   January 21 at   22ʰ
February 6 at 21ʰ   February 21 at   20ʰ

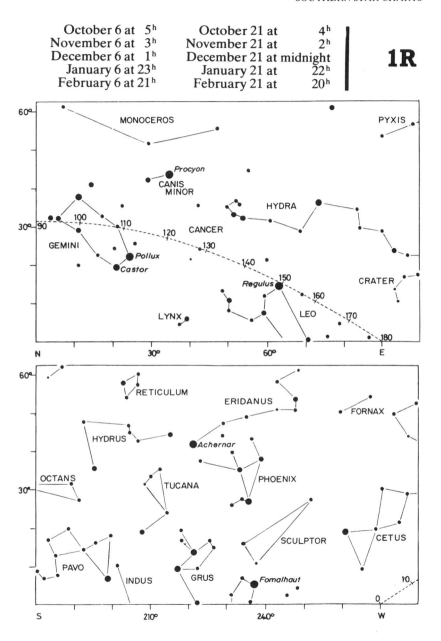

2L

November 6 at 5^h November 21 at 4^h
December 6 at 3^h December 21 at 2^h
January 6 at 1^h January 21 at midnight
February 6 at 23^h February 21 at 22^h
March 6 at 21^h March 21 at 20^h

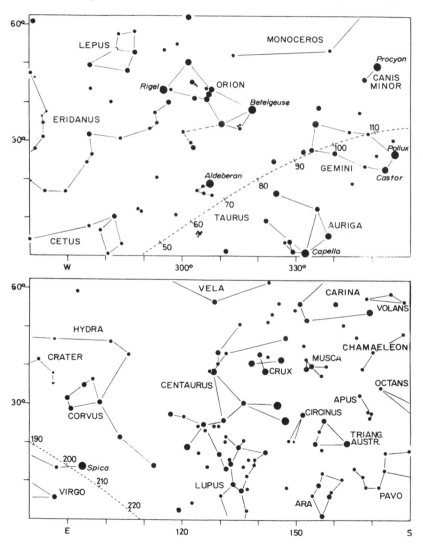

November 6 at 5^h	November 21 at 4^h	
December 6 at 3^h	December 21 at 2^h	
January 6 at 1^h	January 21 at midnight	**2R**
February 6 at 23^h	February 21 at 22^h	
March 6 at 21^h	March 21 at 20^h	

43

3L

January 6 at   3ʰ
February 6 at   1ʰ
March 6 at 23ʰ
April 6 at 21ʰ
May 6 at 19ʰ

January 21 at   2ʰ
February 21 at midnight
March 21 at   22ʰ
April 21 at   20ʰ
May 21 at   18ʰ

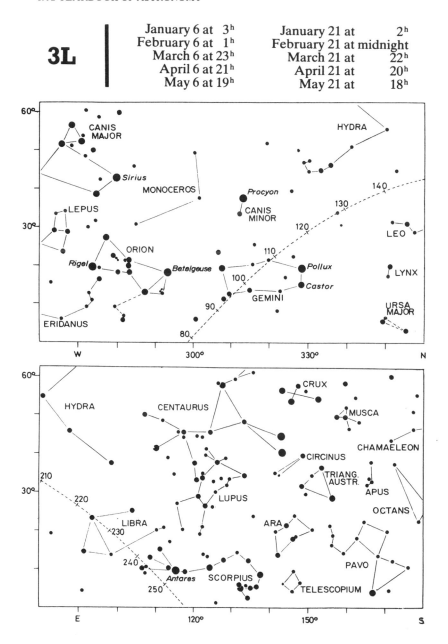

January 6 at 3ʰ January 21 at 2ʰ
February 6 at 1ʰ February 21 at midnight
March 6 at 23ʰ March 21 at 22ʰ
April 6 at 21ʰ April 21 at 20ʰ
May 6 at 19ʰ May 21 at 18ʰ

3R

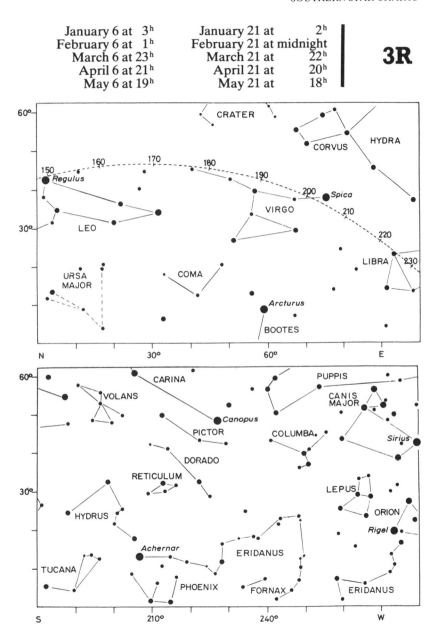

45

4L

February 6 at 3ʰ
March 6 at 1ʰ
April 6 at 23ʰ
May 6 at 21ʰ
June 6 at 19ʰ

February 21 at   2ʰ
March 21 at midnight
April 21 at   22ʰ
May 21 at   20ʰ
June 21 at   18ʰ

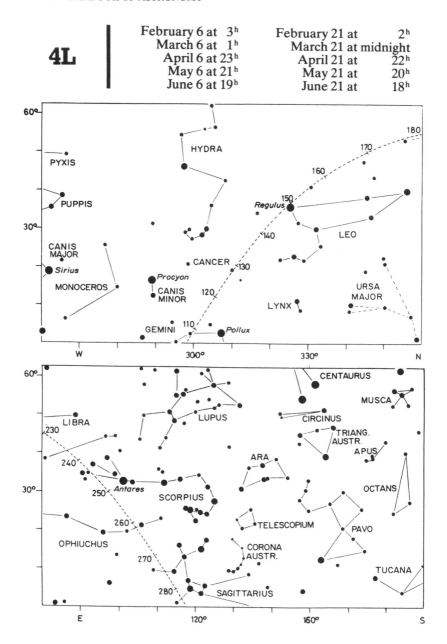

February 6 at 3ʰ	February 21 at 2ʰ
March 6 at 1ʰ	March 21 at midnight
April 6 at 23ʰ	April 21 at 22ʰ
May 6 at 21ʰ	May 21 at 20ʰ
June 6 at 19ʰ	June 21 at 18ʰ

4R

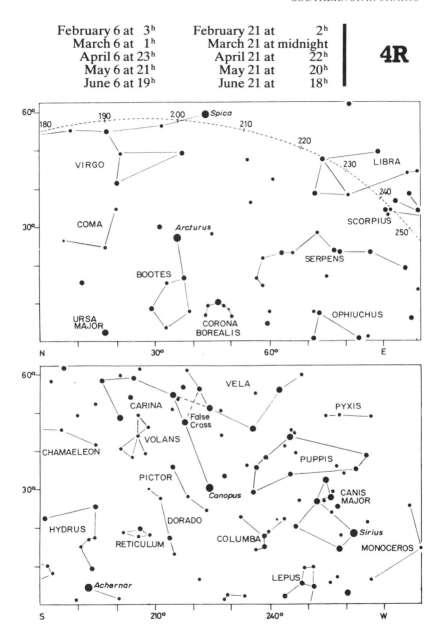

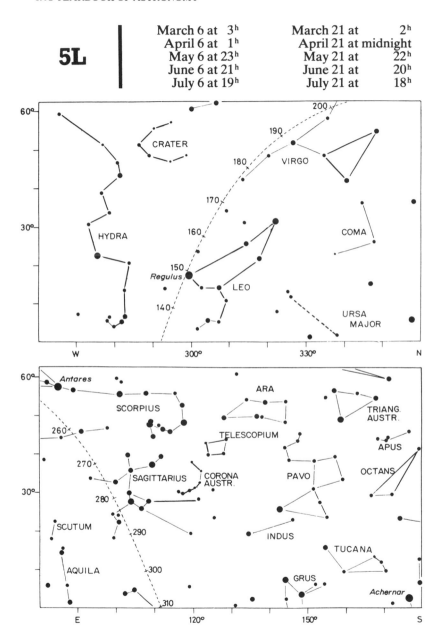

5L

March 6 at 3ʰ	March 21 at 2ʰ
April 6 at 1ʰ	April 21 at midnight
May 6 at 23ʰ	May 21 at 22ʰ
June 6 at 21ʰ	June 21 at 20ʰ
July 6 at 19ʰ	July 21 at 18ʰ

March 6 at 3^h	March 21 at 2^h
April 6 at 1^h	April 21 at midnight
May 6 at 23^h	May 21 at 22^h
June 6 at 21^h	June 21 at 20^h
July 6 at 19^h	July 21 at 18^h

5R

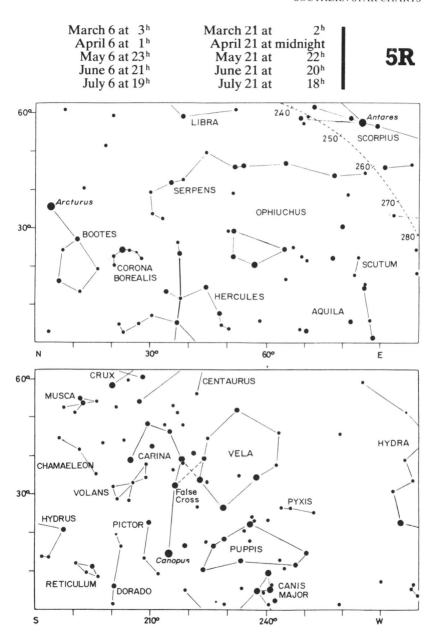

60°

LIBRA   240

Antares

SCORPIUS

250

260

SERPENS

Arcturus   OPHIUCHUS   270

30°

BOOTES   280

CORONA
BOREALIS

SCUTUM

HERCULES

AQUILA

N   30°   60°   E

60°

CRUX   CENTAURUS

MUSCA

HYDRA

CHAMAELEON   CARINA   VELA

30°

VOLANS   False
Cross

PYXIS

HYDRUS   PICTOR

PUPPIS

Canopus

RETICULUM   CANIS
MAJOR

DORADO

S   210°   240°   W

49

6L

March 6 at 5^h March 21 at 4^h
April 6 at 3^h April 21 at 2^h
May 6 at 1^h May 21 at midnight
June 6 at 23^h June 21 at 22^h
July 6 at 21^h July 21 at 20^h

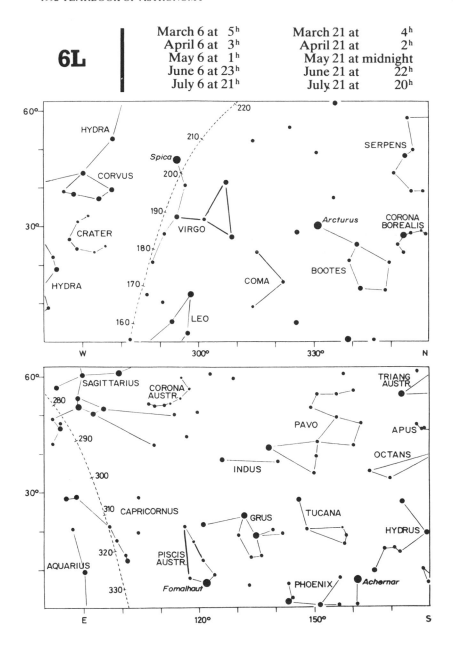

March 6 at   5ʰ   March 21 at   4ʰ
April 6 at   3ʰ   April 21 at   2ʰ
May 6 at   1ʰ   May 21 at midnight   **6R**
June 6 at 23ʰ   June 21 at   22ʰ
July 6 at 21ʰ   July 21 at   20ʰ

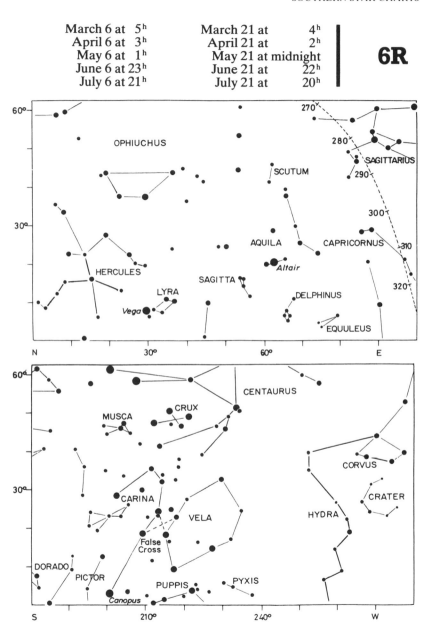

7L

April 6 at 5ʰ
May 6 at 3ʰ
June 6 at 1ʰ
July 6 at 23ʰ
August 6 at 21ʰ

April 21 at 4ʰ
May 21 at 2ʰ
June 21 at midnight
July 21 at 22ʰ
August 21 at 20ʰ

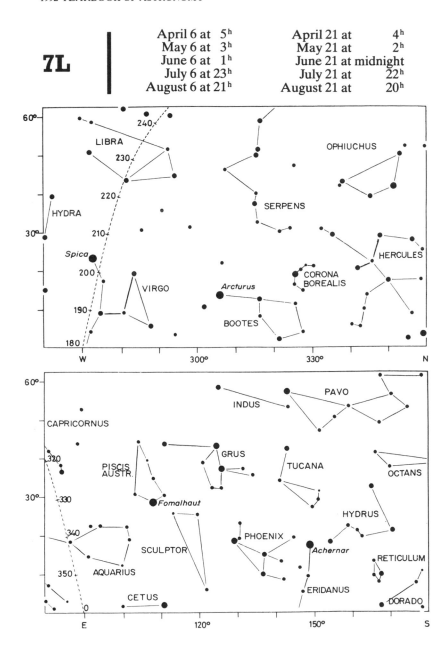

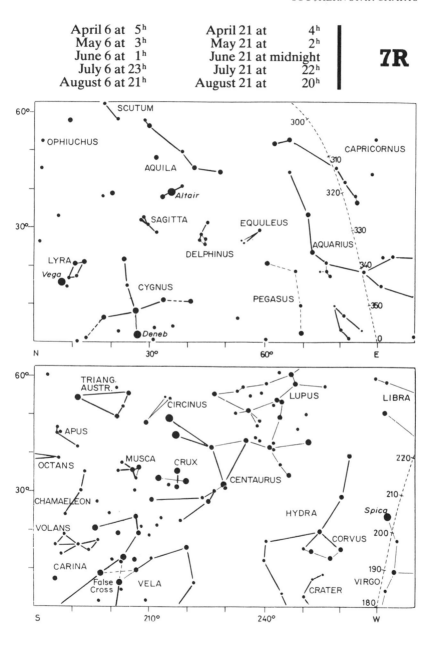

April 6 at 5ʰ	April 21 at 4ʰ	
April 6 at 5ʰ	April 21 at 4ʰ	
May 6 at 3ʰ	May 21 at 2ʰ	
June 6 at 1ʰ	June 21 at midnight	**7R**
July 6 at 23ʰ	July 21 at 22ʰ	
August 6 at 21ʰ	August 21 at 20ʰ	

8L

May 6 at 5^h	May 21 at 4^h
June 6 at 3^h	June 21 at 2^h
July 6 at 1^h	July 21 at midnight
August 6 at 23^h	August 21 at 22^h
September 6 at 21^h	September 21 at 20^h

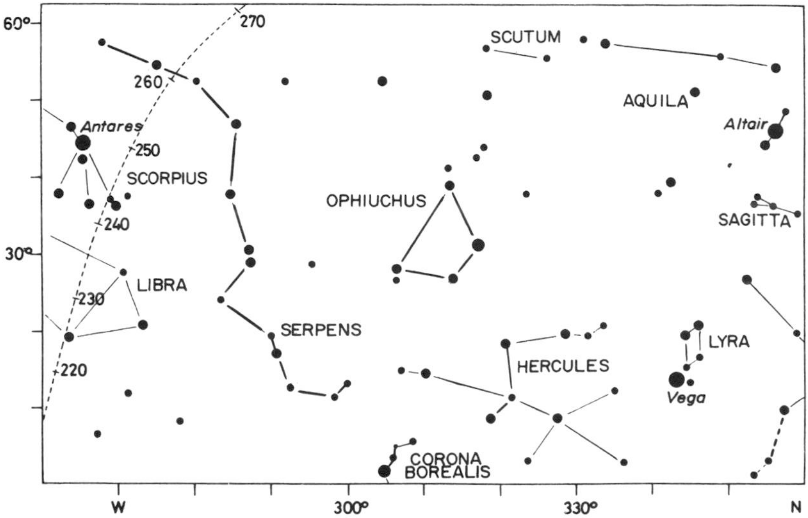

60°

270

SCUTUM

260

AQUILA

Antares

Altair

250

SCORPIUS

OPHIUCHUS

SAGITTA

240

30°

LIBRA

230

SERPENS

LYRA

HERCULES

220

Vega

CORONA
BOREALIS

W   300°   330°   N

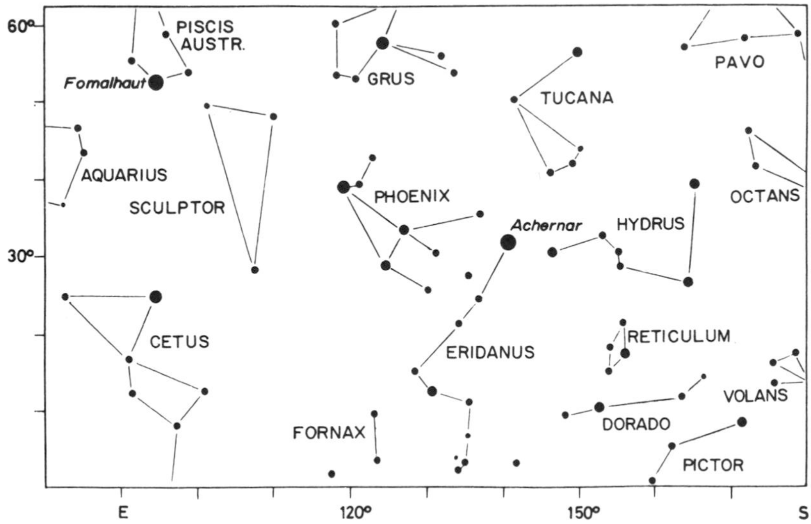

60°

PISCIS
AUSTR.

PAVO

Fomalhaut

GRUS

TUCANA

AQUARIUS

SCULPTOR

PHOENIX

OCTANS

Achernar

HYDRUS

30°

RETICULUM

CETUS

ERIDANUS

VOLANS

DORADO

FORNAX

PICTOR

E   120°   150°   S

May 6 at 5ʰ   May 21 at   4ʰ
June 6 at 3ʰ   June 21 at   2ʰ
July 6 at 1ʰ   July 21 at midnight
August 6 at 23ʰ August 21 at   22ʰ
September 6 at 21ʰ September 21 at 20ʰ

8R

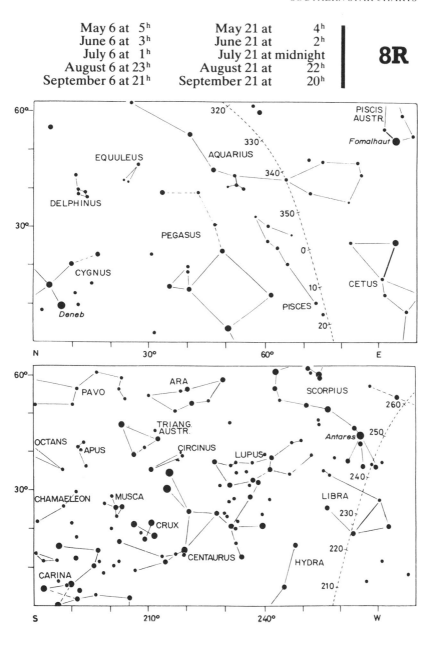

9L

June 6 at 5ʰ	June 21 at 4ʰ
July 6 at 3ʰ	July 21 at 2ʰ
August 6 at 1ʰ	August 21 at midnight
September 6 at 23ʰ	September 21 at 22ʰ
October 6 at 21ʰ	October 21 at 20ʰ

| June 6 at 5ʰ | June 21 at 4ʰ | |
| June 6 at 5ʰ | June 21 at 4ʰ | **9R** |

June 6 at   5ʰ   June 21 at   4ʰ
July 6 at   3ʰ   July 21 at   2ʰ
August 6 at   1ʰ   August 21 at midnight   **9R**
September 6 at 23ʰ   September 21 at   22ʰ
October 6 at 21ʰ   October 21 at   20ʰ

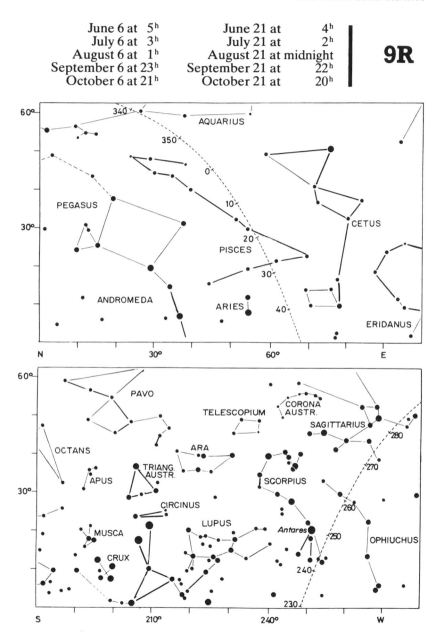

10L

July 6 at 5ʰ	July 21 at 4ʰ
August 6 at 3ʰ	August 21 at 2ʰ
September 6 at 1ʰ	September 21 at midnight
October 6 at 23ʰ	October 21 at 22ʰ
November 6 at 21ʰ	November 21 at 20ʰ

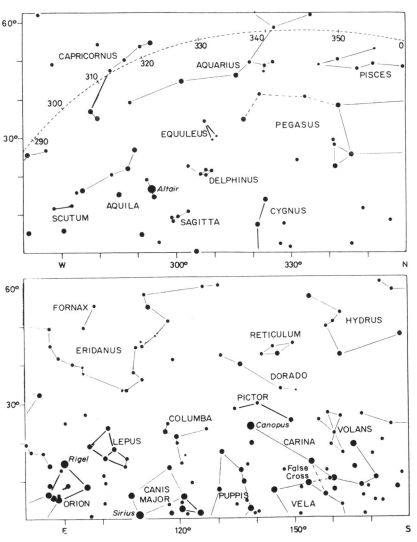

July 6 at 5ʰ	July 21 at 4ʰ	
August 6 at 3ʰ	August 21 at 2ʰ	
September 6 at 1ʰ	September 21 at midnight	**10R**
October 6 at 23ʰ	October 21 at 22ʰ	
November 6 at 21ʰ	November 21 at 20ʰ	

11L

August 6 at 5ʰ	August 21 at 4ʰ
September 6 at 3ʰ	September 21 at 2ʰ
October 6 at 1ʰ	October 21 at midnight
November 6 at 23ʰ	November 21 at 22ʰ
December 6 at 21ʰ	December 21 at 20ʰ

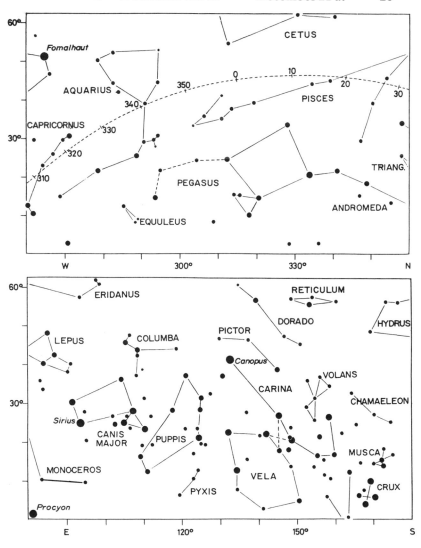

August 6 at 5ʰ	August 21 at 4ʰ	
September 6 at 3ʰ	September 21 at 2ʰ	
October 6 at 1ʰ	October 21 at midnight	**11R**
November 6 at 23ʰ	November 21 at 22ʰ	
December 6 at 21ʰ	December 21 at 20ʰ	

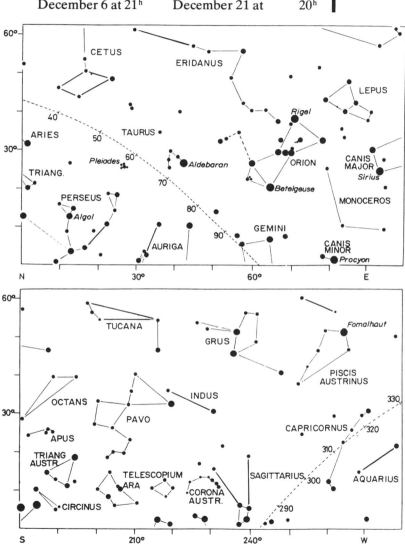

12L

September 6 at 5ʰ	September 21 at 4ʰ
October 6 at 3ʰ	October 21 at 2ʰ
November 6 at 1ʰ	November 21 at midnight
December 6 at 23ʰ	December 21 at 22ʰ
January 6 at 21ʰ	January 21 at 20ʰ

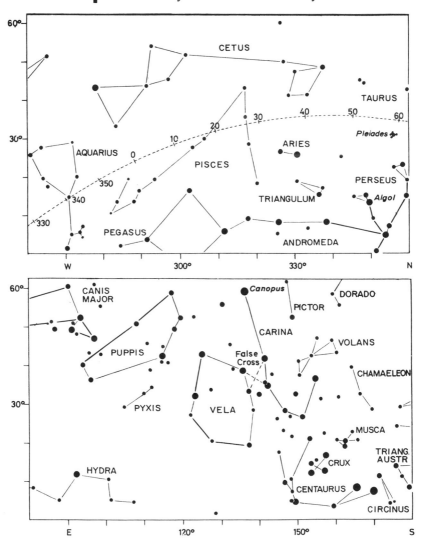

September 6 at 5^h	September 21 at 4^h	
October 6 at 3^h	October 21 at 2^h	**12R**
November 6 at 1^h	November 21 at midnight	
December 6 at 23^h	December 21 at 22^h	
January 6 at 21^h	January 21 at 20^h	

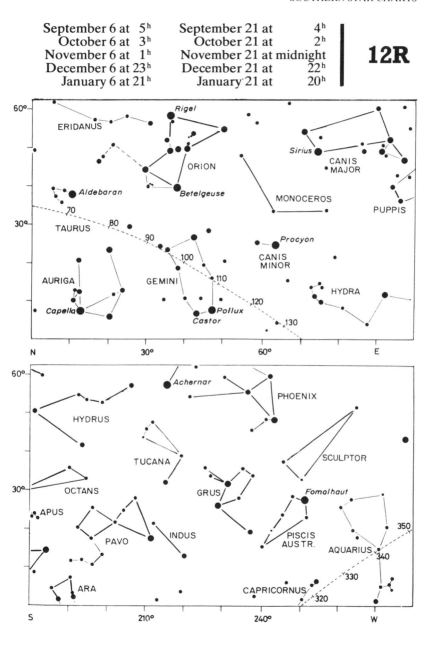

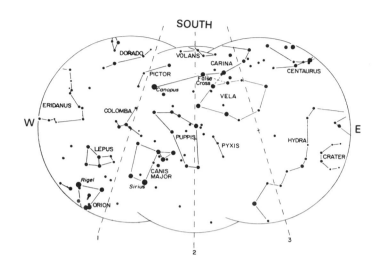

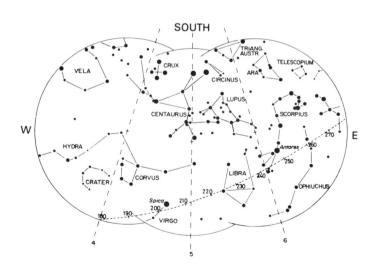

Southern Hemisphere Overhead Stars

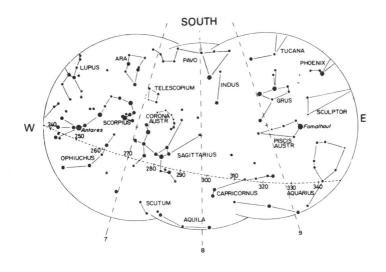

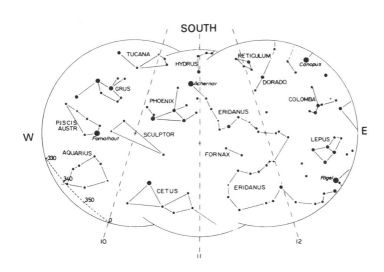

Southern Hemisphere Overhead Stars

The Planets and the Ecliptic

The paths of the planets about the Sun all lie close to the plane of the ecliptic, which is marked for us in the sky by the apparent path of the Sun among the stars, and is shown on the star charts by a broken line. The Moon and planets will always be found close to this line, never departing from it by more than about 7 degrees. Thus the planets are most favourably placed for observation when the ecliptic is well displayed, and this means that it should be as high in the sky as possible. This avoids the difficulty of finding a clear horizon, and also overcomes the problem of atmospheric absorption, which greatly reduces the light of the stars. Thus a star at an altitude of 10 degrees suffers a loss of 60 per cent of its light, which corresponds to a whole magnitude; at an altitude of only 4 degrees, the loss may amount to two magnitudes.

The position of the ecliptic in the sky is therefore of great importance, and since it is tilted at about 23½ degrees to the Equator, it is only at certain times of the day or year that it is displayed to the best advantage. It will be realized that the Sun (and therefore the ecliptic) is at its highest in the sky at noon in midsummer, and at its lowest at noon in midwinter. Allowing for the daily motion of the sky, these times lead to the fact that the ecliptic is highest at midnight in winter, at sunset in the spring, at noon in summer and at sunrise in the autumn. Hence these are the best times to see the planets. Thus, if Venus is an evening object, in the western sky after sunset, it will be seen to best advantage if this occurs in the spring, when the ecliptic is high in the sky and slopes down steeply to the horizon. This means that the planet is not only higher in the sky, but will remain for a much longer period above the horizon. For similar reasons, a morning object will be seen at its best on autumn mornings before sunrise, when the ecliptic is high in the east. The outer planets, which can come to opposition (i.e. opposite the Sun), are best seen when opposition occurs in the winter months, when the ecliptic is high in the sky at midnight.

The seasons are reversed in the Southern Hemisphere, spring beginning at the September Equinox, when the Sun crosses the Equator on its way south, summer beginning at the December

Solstice, when the Sun is highest in the southern sky, and so on. Thus, the times when the ecliptic is highest in the sky, and therefore best placed for observing the planets, may be summarized as follows:

	Midnight	*Sunrise*	*Noon*	*Sunset*
Northern lats.	December	September	June	March
Southern lats.	June	March	December	September

In addition to the daily rotation of the celestial sphere from east to west, the planets have a motion of their own among the stars. The apparent movement is generally *direct*, i.e. to the east, in the direction of increasing longitude, but for a certain period (which depends on the distance of the planet) this apparent motion is reversed. With the outer planets this *retrograde* motion occurs about the time of opposition. Owing to the different inclination of the orbits of these planets, the actual effect is to cause the apparent path to form a loop, or sometimes an S-shaped curve. The same effect is present in the motion of the inferior planets, Mercury and Venus, but it is not so obvious, since it always occurs at the time of inferior conjunction.

The inferior planets, Mercury and Venus, move in smaller orbits than that of the Earth, and so are always seen near the Sun. They are most obvious at the times of greatest angular distance from the Sun (greatest elongation), which may reach 28 degrees for Mercury, or 47 degrees for Venus. They are seen as evening objects in the western sky after sunset (at eastern elongations) or as morning objects in the eastern sky before sunrise (at western elongations). The succession of phenomena, conjunctions and elongations, always follows the same order, but the intervals between them are not equal. Thus, if either planet is moving round the far side of its orbit its motion will be to the east, in the same direction in which the Sun appears to be moving. It therefore takes much longer for the planet to overtake the Sun – that is, to come to superior conjunction – than it does when moving round to inferior conjunction, between Sun and Earth. The intervals given in the following table are average values; they remain fairly constant in the case of Venus, which travels in an almost circular orbit. In the case of Mercury, however, conditions vary widely because of the great eccentricity and inclination of the planet's orbit.

		Mercury	*Venus*
Inferior conj.	to Elongation West	22 days	72 days
Elongation West	to Superior conj.	36 days	220 days
Superior conj.	to Elongation East	36 days	220 days
Elongation East	to Inferior conj.	22 days	72 days

The greatest brilliancy of Venus always occurs about 36 days before or after inferior conjunction. This will be about a month *after* greatest eastern elongation (as an evening object), or a month *before* greatest western elongation (as a morning object). No such rule can be given for Mercury, because its distance from the Earth and the Sun can vary over a wide range.

Mercury is not likely to be seen unless a clear horizon is available. It is seldom seen as much as 10 degrees above the horizon in the twilight sky in northern latitudes, but this figure is often exceeded in the Southern Hemisphere. This favourable condition arises because the maximum elongation of 28 degrees can occur only when the planet is at aphelion (farthest from the Sun), and this point lies well south of the Equator. Northern observers must be content with smaller elongations, which may be as little as 18 degrees at perihelion. In general, it may be said that the most favourable times for seeing Mercury as an evening object will be in spring, some days before greatest eastern elongation; in autumn, it may be seen as a morning object some days after greatest western elongation.

Venus is the brightest of the planets and may be seen on occasions in broad daylight. Like Mercury, it is alternately a morning and an evening object, and it will be highest in the sky when it is a morning object in autumn, or an evening object in spring. The phenomena of Venus given in the table above can occur only in the months of January, April, June, August and November, and it will be realized that they do not all lead to favourable apparitions of the planet. In fact, Venus is to be seen at its best as an evening object in northern latitudes when eastern elongation occurs in June. The planet is then well north of the Sun in the preceding spring months, and is a brilliant object in the evening sky over a long period. In the Southern Hemisphere a November elongation is best. For similar reasons, Venus gives a prolonged display as a morning object in the months following western elongation in November (in northern latitudes) or in June (in the Southern Hemisphere).

The superior planets, which travel in orbits larger than that of the Earth, differ from Mercury and Venus in that they can be seen opposite the Sun in the sky. The superior planets are morning objects after conjunction with the Sun, rising earlier each day until they come to opposition. They will then be nearest to the Earth (and therefore at their brightest), and will then be on the meridian at midnight, due south in northern latitudes, but due north in the Southern Hemisphere. After opposition they are evening objects, setting earlier each evening until they set in the west with the Sun at the next conjunction. The change in brightness about the time of opposition is most noticeable in the case of Mars, whose distance from Earth can vary considerably and rapidly. The other superior planets are at such great distances that there is very little change in brightness from one opposition to another. The effect of altitude is, however, of some importance, for at a December opposition in northern latitudes the planets will be among the stars of Taurus or Gemini, and can then be at an altitude of more than 60 degrees in southern England. At a summer opposition, when the planet is in Sagittarius, it may only rise to about 15 degrees above the southern horizon, and so makes a less impressive appearance. In the Southern Hemisphere, the reverse conditions apply; a June opposition being the best, with the planet in Sagittarius at an altitude which can reach 80 degrees above the northern horizon for observers in South Africa.

Mars, whose orbit is appreciably eccentric, comes nearest to the Earth at an opposition at the end of August. It may then be brighter even than Jupiter, but rather low in the sky in Aquarius for northern observers, though very well placed for those in southern latitudes. These favourable oppositions occur every fifteen or seventeen years (1956, 1971, 1988, 2003) but in the Northern Hemisphere the planet is probably better seen at an opposition in the autumn or winter months, when it is higher in the sky. Oppositions of Mars occur at an average interval of 780 days, and during this time the planet makes a complete circuit of the sky.

Jupiter is always a bright planet, and comes to opposition a month later each year, having moved, roughly speaking, from one Zodiacal constellation to the next.

Saturn moves much more slowly than Jupiter, and may remain in the same constellation for several years. The brightness of Saturn depends on the aspects of its rings, as well as on the distance from Earth and Sun. The rings were inclined towards the Earth and Sun in 1980 and are currently near their maximum opening. The next passage of both Earth and Sun through the ring-plane will not occur until 1995.

Uranus, *Neptune*, and *Pluto* are hardly likely to attract the attention of observers without adequate instruments.

Phases of the Moon
1992

	New Moon				First Quarter				Full Moon				Last Quarter		
	d	h	m		d	h	m		d	h	m		d	h	m
Jan.	4	23	10	Jan.	13	02	32	Jan.	19	21	28	Jan.	26	15	27
Feb.	3	19	00	Feb.	11	16	15	Feb.	18	08	04	Feb.	25	07	56
Mar.	4	13	22	Mar.	12	02	36	Mar.	18	18	18	Mar.	26	02	30
Apr.	3	05	01	Apr.	10	10	06	Apr.	17	04	42	Apr.	24	21	40
May	2	17	44	May	9	15	43	May	16	16	03	May	24	15	53
June	1	03	57	June	7	20	47	June	15	04	50	June	23	08	11
	30	12	18												
July	29	19	35	July	7	02	43	July	14	19	06	July	22	22	12
Aug.	28	02	42	Aug.	5	10	58	Aug.	13	10	27	Aug.	21	10	01
Sept.	26	10	40	Sept.	3	22	39	Sept.	12	02	17	Sept.	19	19	53
Oct.	25	20	34	Oct.	3	14	12	Oct.	11	18	03	Oct.	19	04	12
Nov.	24	09	11	Nov.	2	09	11	Nov.	10	09	20	Nov.	17	11	39
Dec.	24	00	43	Dec.	2	06	17	Dec.	9	23	41	Dec.	16	19	13

All times are G.M.T.
Reproduced, with permission, from data supplied by the Science and Engineering Research Council.

Longitudes of the Sun, Moon and Planets in 1992

DATE		Sun	Moon	Venus	Mars	Jupiter	Saturn
		°	°	°	°	°	°
January	6	285	296	246	267	165	306
	21	300	136	264	278	164	308
February	6	316	341	284	291	163	310
	21	332	188	302	302	161	312
March	6	346	2	320	313	159	313
	21	1	210	338	324	157	315
April	6	16	50	358	337	156	316
	21	31	258	16	348	155	317
May	6	46	88	35	0	155	318
	21	60	290	53	11	155	318
June	6	76	141	73	24	157	318
	21	90	334	91	34	158	318
July	6	104	180	110	45	161	317
	21	118	6	128	56	163	316
August	6	134	230	148	67	166	315
	21	148	53	166	76	169	314
September	6	164	277	187	86	173	313
	21	178	103	204	95	176	312
October	6	193	309	223	103	179	312
	21	208	142	241	109	182	312
November	6	224	353	261	114	185	312
	21	239	195	278	117	188	313
December	6	254	26	297	117	190	314
	21	269	233	314	114	192	315

Longitude of *Uranus* 286°   *Moon:* Longitude of ascending node
　　　　　　　Neptune 288°　　　　Jan. 1: 280°　　　Dec. 31: 260°

Mercury moves so quickly among the stars that it is not possible to indicate its position on the star charts at a convenient interval. The monthly notes must be consulted for the best times at which the planet may be seen.

The positions of the other planets are given in the table on the previous page. This gives the apparent longitudes on dates which correspond to those of the star charts, and the position of the planet may at once be found near the ecliptic at the given longitude.

Examples
In the Southern Hemisphere two planets are seen in the eastern morning sky in late March. Identify them.

The southern star chart 6R shows the eastern sky at March 21^{d}4^h and shows longitudes 270°–325°. Reference to the table on page 71 gives the longitude of Mars as 324° and that of Saturn as 315°, on March 21. Thus these planets are found low in the eastern sky and the one with the slightly reddish tint is Mars.

The positions of the Sun and Moon can be plotted on the star maps in the same manner as for the planets. The average daily motion of the Sun is 1°, and of the Moon 13°. For the Moon an indication of its position relative to the ecliptic may be obtained from a consideration of its longitude relative to that of the ascending node. The latter changes only slowly during the year as will be seen from the values given on the previous page. Let us call the difference in longitude of Moon-node, d. Then if d = 0°, 180° or 360° the Moon is on the ecliptic. If d = 90° the Moon is 5° north of the ecliptic and if d = 270° the Moon is 5° south of the ecliptic.

On July 6 the Moon's longitude is given as 180° and the longitude of the node is found by interpolation to be about 272°. Thus d = 268° and the Moon is about 5° south of the ecliptic. Its position may be plotted on northern star charts 2L, 3L, 5R and 6R: and southern star charts 2R, 3R, 5L and 6L.

Events in 1992

ECLIPSES

There will be five eclipses, three of the Sun and two of the Moon.

January 4–5: annular eclipse of the Sun – west coast of North America.
June 15: partial eclipse of the Moon – Africa, America, New Zealand.
June 30: total eclipse of the Sun – South America.
December 9–10: total eclipse of the Moon – Asia, Europe, Africa, America.
December 23–24: partial eclipse of the Sun – East Asia.

THE PLANETS

Mercury may be seen more easily from northern latitudes in the evenings about the time of greatest eastern elongation (March 9) and in the mornings around greatest western elongation (December 9). In the Southern Hemisphere the dates are April 23 (morning) and October 31 (evening).

Venus is visible in the mornings until April and in the evenings from August until the end of the year.

Mars is visible in the mornings throughout the year.

Jupiter is at opposition on February 29.

Saturn is at opposition on August 7.

Uranus is at opposition on July 7.

Neptune is at opposition on July 9.

Pluto is at opposition on May 12.

JANUARY

New Moon: January 4   *Full Moon:* January 19

EARTH is at perihelion (nearest to the Sun) on January 3 at a distance of 147 million kilometres.

MERCURY is a morning object, magnitude −0.3, at the beginning of the month, visible low in the south-eastern sky at the time of beginning of morning civil twilight. As the planet is well south of the Equator it is best seen from the Southern Hemisphere where its period of visibility lasts for the first three weeks of January. The length of this period decreases the further north the observer is situated so that for observers in the latitude of the British Isles, Mercury can only be glimpsed for the first few days of the month.

VENUS, like Mercury, is also visible in the morning skies, though much further from the Sun than the latter. Venus has a magnitude of −4.0.

MARS is very gradually becoming visible as a morning object, magnitude +1.5, though not to observers in northern temperate latitudes.

JUPITER is a brilliant morning object, magnitude −2.3. It becomes visible in the eastern sky well before midnight and is west of the meridian long before dawn. Figure 1 shows its path amongst the stars throughout the year.

SATURN is too close to the Sun for observation by those in the latitudes of the British Isles. For observers in equatorial and southern latitudes Saturn is a difficult evening object low in the south-western sky for a short time after dark – and then only for the first week or ten days of the month. Saturn is in conjunction with the Sun on January 29.

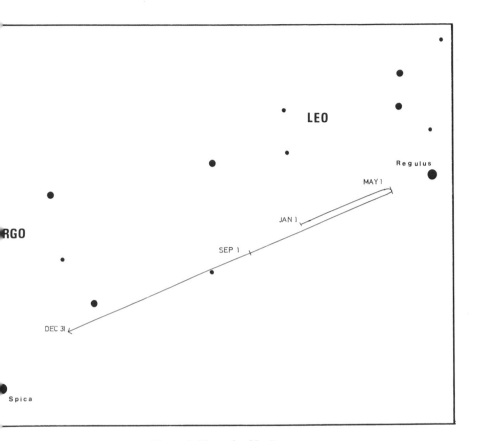

Figure 1. The path of Jupiter.

THE ASTRONOMICAL UNIT. The Earth's orbit round the Sun is not circular; it is an ellipse, with an eccentricity of 0.017. The changing distance from the Sun has only a minor effect upon the seasons, which are due mainly to the axial inclination of 23½°, though some Europeans find it hard to credit that we are actually at our closest to the Sun in early January!

The mean distance from the Sun is known as the *astronomical unit*. It is equal to 149,597,900 km, or 92.9 million miles. The first reasonably good measurement was made by G. D. Cassini in 1672; earlier estimates were wide of the mark. It is interesting to look back at some of them:

Year	Astronomer	Estimated Earth/Sun distance, miles
BC		
270	Aristarchus	3,000,000
130	Hipparchus	3,000,000
AD		
140	Ptolemy	5,000,000
1543	Copernicus	2,000,000
1600	Tycho Brahe	5,000,000
1618	Kepler	14,000,000
1672	Cassini	86,000,000
1770	Euler	92,897,000

All estimates since then have been at least reasonably good. The present value is accurate to a tiny fraction of one per cent.

TWO GREAT ASTRONOMERS. January is the centenary of the deaths of two of the greatest Victorian astronomers, who died within three weeks of each other: John Couch Adams on January 21, 1892, and Sir George Biddell Airy on January 2, 1892. This is ironical in view of the fact that their names were linked in one of the most famous of all unfortunate episodes in the history of astronomy – the failure to discover the planet Neptune.

Adams was born at Lidcot, in Cornwall, on June 5, 1819. He was educated at a private school in Devonport, and then went to Cambridge University, graduating in 1843 with top honours. Even before taking his degree he had made up his mind to investigate the movements of the planet Uranus, which had been discovered by William Herschel in 1781 but was not moving as it had been expected to do. Adams thought, quite correctly, that its irregular motion was due to the perturbing effect of a more remote planet, and he believed that he could compute the position of the planet responsible.

By 1845 he had worked out a position for the planet, and submitted his results to the Astronomer Royal, George Biddell Airy, at Greenwich. Airy, born at Alnwick in Northumberland on July 27, 1801, had graduated from Cambridge in 1823 and had been Professor of Astronomy there until 1835, when he was called to take up his appointment at Greenwich. He was unquestionably a great astronomer, who modernized the Observatory and became a leader in world science. Unfortunately he did not take immediate action when learning of Adams's work. He did finally organize a search, which was carried out at Cambridge by James Challis with the Northumberland refractor (there was no suitable telescope at Greenwich at that time), but it was too late; while Challis was still searching, the planet was identified from Berlin by Johann Galle and Heinrich D'Arrest on the basis of similar calculations by the French mathematician Urbain Le Verrier.

When the French learned that Adams had made accurate calculations well before Le Verrier had done so, they assumed that the English were trying to claim priority for the discovery of the planet we now call Neptune. Feelings ran high, and almost created an international incident, but fortunately neither Adams nor Le Verrier took part – and when they met face to face, some time later, they struck up an immediate friendship, even though Adams could not speak French and Le Verrier was equally unfamiliar with English.

Airy was bitterly blamed for the failure to detect Neptune. It is true that all the information had been put into his hands, but there were extenuating circumstances, and moreover it emerged that Challis had observed Neptune during the first week of his search – and would have recognized it as a planet if only he had compared his observations instead of putting them aside.

Airy's régime at Greenwich lasted until 1881. When he retired Adams was invited to succeed him as Astronomer Royal, but declined; in 1858 he had succeeded Challis as Professor of Astronomy at Cambridge and Director of the Observatory there.

It is probably true to say that Airy is remembered nowadays mainly for his role in the Neptune episode. This is quite unfair. He accomplished so much for astronomy, quite apart from his re-organization of Greenwich Observatory, that there is no justification for harping upon the one minor blot upon his career.

FEBRUARY

New Moon: February 3   *Full Moon:* February 18

MERCURY reaches superior conjunction on February 12 and is therefore too close to the Sun for observation for most of the month. However, for the last few evenings in February it may be seen low above the western horizon, magnitude −1, at the time of end of evening civil twilight.

VENUS is a brilliant object in the morning skies, magnitude −4.0. Being well south of the Equator the planet will only be seen by observers in northern temperate latitudes for a short period before dawn, low above the south-eastern horizon.

MARS is a morning object, visible low above the south-eastern horizon about an hour before sunrise, though not for observers in northern temperate latitudes. Mars passes 0°.8 S of Venus on February 19.

JUPITER, magnitude −2.5, continues to be visible as a brilliant object in the night sky. It is at opposition on the last day of the month and is therefore visible throughout the hours of darkness. At opposition Jupiter is 660 million kilometres from the Earth.

SATURN, magnitude +0.8, is emerging slowly from the morning twilight, though observers as far north as the latitudes of the British Isles will have to wait until April before they will be able to see it. Observers further south should be able to detect it low above the south-eastern horizon for a short while before the morning twilight gets too strong, during the last week of the month. Do not confuse Mars with Saturn. Saturn is slightly brighter and also closer to the Sun than Mars.

THE CELESTIAL LYNX. It is all too obvious that the constellations we recognize today are very unequal in size and importance (no less a person than Sir John Herschel once commented that they seem to have been designed so as to cause as much confusion and inconvenience as possible). One of the more obscure groups is Lynx, the Lynx, which was added to the sky by Hevelius in 1690. It is in the far north, and during February evenings it is very high for observers in Britain or the northern States.

Lynx lies mainly between Ursa Major and Gemini. It has no particular pattern; the only brightish star (Alpha Lyncis) is of magnitude 3.1, and forms an equilateral triangle with Pollux in Gemini and Regulus in Leo. It has an M-type spectrum, and is obviously red when seen with binoculars. The Mira-type variable R Lyncis can reach magnitude 7.2, and is of type S, so that it is very red; but all in all Lynx may be classed as one of the most barren regions of the sky. It has even been said that Hevelius created it because an observer needs to be lynx-eyed to see anything there at all!

HALLEY'S COMET. A year ago, in February 1991, a most remarkable thing happened to Halley's Comet. As many people will know, the comet takes seventy-six years to make one journey round the Sun; it was back in 1910 and 1986, and will be again at perihelion in 2061. During the 1986 apparition it was much less brilliant than usual, simply because it was so badly placed from our point of view; when the comet reached perihelion, the Earth was on the other side of the Sun. Though the comet became an easy naked-eye object, it was certainly not spectacular.

As it drew away from the Sun, it faded, and the tail disappeared. Over the next few years it was followed at major observatories, notably by astronomers at La Silla, in Chile, using the 1.5-metre Danish reflector there. It seemed that the comet had gone into 'hibernation' – as all comets do when far from the Sun – but then, in February 1991, there was an outburst which produced a cloud of dust round the icy nucleus. At that time the comet was well beyond the orbit of Saturn, with an estimated temperature of −200 degrees Centigrade.

Nothing of the sort had been expected, and it was very hard to explain. Had the comet been hit by an errant asteroid? Had there been some delayed explosion in the nucleus, or was solar radiation responsible? All these three explanations seemed to be very improbable, and at the time when these words are being written we do not pretend to know the answer. Unfortunately the comet is receding all the time, and before long we will inevitably lose sight of it until it draws inward once more.

TWO CENTENARIES. There are two centenaries this month – of very different astronomers: one a distinguished professional, the other a distinguished amateur. They did, however, have one link. Both were students of variable stars.

Cuno Hoffmeister was born on February 2, 1892 at Sonneberg, in Germany. He acted as assistant at the Bamberg and Jena Observatories, and then returned home to set up his own observatory. By 1925 it was well-equipped, and three years later he began to collaborate with the Berlin-Babelsberg Observatory in a photographic sky patrol, mainly with the object of detecting new variable stars above the twelfth magnitude.

Over the years Hoffmeister turned his private observatory into a world centre for variable star research; the telescopes included a 24-inch Cassegrain, a fast double Schmidt, and a 16-inch photographic refractor. He personally found over 10,000 new variables, and by 1986, when he died, over a hundred thousand sky patrol plates had been taken. Hoffmeister was also interested in meteors (his book *Meteoric Currents* appeared in 1948) and in the Zodiacal Light, which he studied with a visual photometer of his own invention. For many years he was editor of the German periodical *Die Sterne*.

Henden Egerton Houghton, born in London on February 28, 1892, spent almost all his life in South Africa. When he arrived in Cape Town he joined the local amateur group there, and became director of its variable star group upon the retirement of G. E. Ensor in 1934. He became President of the Astronomical Society of South Africa in 1937, and in his presidential address he said that he had made some 20,800 observations of 90 stars; he also made an independent discovery of a comet (1932b, Houghton-Ensor). Apart from his astronomical activities, he gave many years' service to the Office of the High Commissioner, and was awarded the MBE in 1937. He died in 1947.

MARCH

New Moon: March 4   *Full Moon:* March 18

Summer Time in Great Britain and Northern Ireland commences on March 29.

Equinox: March 20

MERCURY is visible as an evening object. For observers in northern temperate latitudes this will be the most favourable evening apparition of the year. Figure 2 shows, for observers in latitude N.52°, the changes in azimuth (true bearing from the north through east, south, and west) and altitude of Mercury on successive evenings when the Sun is 6° below the horizon. This condition is known as the end of evening civil twilight, and in this latitude and at this time of year occurs about 35 minutes after sunset. The changes in the brightness of the planet are indicated by the relative sizes of the circles marking Mercury's position at five-day intervals. It will be noticed that Mercury is at its brightest before it reaches greatest eastern elongation (18°) on March 9. The magnitude of Mercury fades from −1.1 on March 1 to +2.1 on March 19. For the last ten days of the month it is too close to the Sun for observation.

VENUS continues to be visible as a brilliant morning object with a magnitude of −3.9. It is slowly drawing in towards the Sun and observers in the latitudes of the British Isles will be finding it increasingly difficult to observe as the month progresses; and by the end of March Venus will be rising at about the same time as the Sun.

MARS continues to be visible as a morning object for observers in the tropics and in southern latitudes.

JUPITER is a brilliant object in Leo and since opposition was on

February 28, continues to be visible for the greater part of the night. Its magnitude is −2.5.

SATURN is still unsuitably placed for observation by those in the latitudes of the British Isles. Those further south can see the planet as a morning object, magnitude +0.8, low in the south-eastern sky, before it fades in the morning twilight. Early in the month Saturn passes 0°.4N. of Mars and is then further from the Sun. Saturn is slightly brighter than Mars.

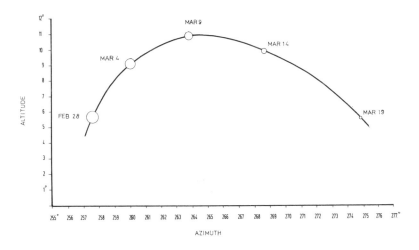

Figure 2. Evening apparition of Mercury for latitude N.52°.

THE STRANGE CALENDAR OF MERCURY. No Earth-based telescope will show much detail on the surface of Mercury. The planet is always inconveniently close to the Sun in the sky; it is small, and a long way away. Therefore, most of our knowledge of it comes from one space-craft, Mariner 10, which made three active passes of the planet in 1974–5 before its equipment failed.

Up to the 1960s it was thought that Mercury had a captured or synchronous rotation, in which case the 'day' would have been equal to the 'year' of 88 Earth-days, and Mercury would have kept the same face turned sunward all the time (just as the Moon always keeps the same face turned toward the Earth). In fact this is not so. Mercury's rotation period is 58.6 days, or two-thirds of a Mercurian year. Because the orbit is more eccentric than ours – the distance of Mercury from the Sun at perigee is 28,500,000 miles (45,900,000 km) and at apogee as much as 43,300,000 miles (69,700,000 km) the apparent diameter of the Sun changes, and so does its apparent motion in the Mercurian sky. There are two 'hot poles', from which the Sun would be overhead at the time of perigee, and the surface temperature would reach around 800 degrees Fahrenheit (430 degrees Centigrade). One of these is marked by the great Caloris Basin, which is 800 miles across and is bounded by rings of high mountain blocks.

At sunrise over Caloris, the Sun would be at its smallest; as it climbed toward the zenith it would swell, but before reaching the zenith it would 'backtrack' for a period before resuming its normal motion. As it dropped toward the horizon it would shrink, finally setting 88 Earth-days after having risen. The interval between one Mercurian sunrise and the next is 176 Earth-days or two Mercurian years.

To an observer 90° away from Caloris, the Sun would be largest when rising and setting, and there would be a 'false sunrise' and a 'false sunset' as the Sun bobbed above and below the horizon. The reason for this strange behaviour is that when Mercury is near perihelion its orbital speed is very great, whereas the rate of axial spin remains constant.

Because Mercury is the closest planet to the Sun it might be expected to be the hottest, but this is not so; the surface temperature of Venus is appreciably higher – because Venus has a dense, carbon-dioxide rich atmosphere which shuts in the Sun's heat, whereas the atmosphere of Mercury is negligible. We must admit that neither of these inner planets is suited for manned exploration, at least in the foreseeable future.

EQUATORIAL STARS. Orion, one of the most magnificent constellations in the sky, is still well above the horizon during March evenings, and since it is crossed by the celestial equator it can be seen from every inhabited part of the world. The equator passes close to Delta Orionis (Mintaka), the faintest of the three members of the Hunter's Belt; the magnitude is 2.2 (actually it is very slightly variable).

Stars close to the equator can be used to measure the diameter of a telescopic eyepiece. The method is to time the interval taken for the star to pass centrally across the field from one side to the other.

This interval, when expressed in minutes and seconds, is then multiplied by 15, and this gives the diameter of the field in minutes and seconds of arc.

The following list gives the stars brighter than magnitude 4.75 which lie within one degree of the celestial equator:

Star	R.A. (epoch 2000.0)			Declination			Magnitude
	h	m	s	°	′	″	
Zeta Aquarii	22	28	49.5	−00	01	13	4.5
Nu Aquilæ	19	26	30.9	00	22	44	4.7
Delta Ceti	02	39	28.9	00	19	43	4.1
Delta Monocerotis	07	11	51.8	−00	29	34	4.1
41 Ophiuchi	17	16	36.5	−00	26	43	4.7
Delta Orionis	05	32	00.3	−00	17	57	2.2
Omicron Orionis	05	28	45.7	−00	22	57	4.7
Zeta Virginis	13	34	41.5	−00	35	46	3.4
Eta Virginis	12	19	54.3	−00	40	00	3.9

This method is a useful 'trick', and the accuracy is quite adequate.

APRIL

New Moon: April 3   *Full Moon:* April 17

MERCURY becomes a morning object early in the month, though not to observers in northern temperate latitudes. For observers in southern latitudes this will be the most favourable morning apparition of the year. Figure 3 shows, for observers in latitude S.35°, the changes in azimuth (true bearing from the north through east, south, and west) and altitude of Mercury on successive mornings when the Sun is 6° below the horizon. This condition is known as the beginning of morning civil twilight, and in this latitude and at this time of year occurs about 30 minutes before sunrise. The changes in the brightness of the planet are indicated by the relative sizes of the circles marking Mercury's position at five-day intervals. It will be noticed that Mercury is at its brightest after it reaches greatest western elongation (27°) on April 23. Note the proximity to Venus, which is about five magnitudes brighter than Mercury.

VENUS is already lost to view for observers in northern temperate latitudes. It continues to move in towards the Sun and for observers further south it is only visible for a short while before sunrise, low above the eastern horizon.

MARS continues to be visible as a morning object, magnitude +1, though not to observers in the British Isles. Figure 4, given with the notes for May, shows the path of Mars amongst the stars during the first part of the year.

JUPITER, magnitude −2.3, continues to be visible as a brilliant object in the night sky, though it is now setting in the west long before dawn.

SATURN, magnitude +0.8, is a morning object in Capricornus

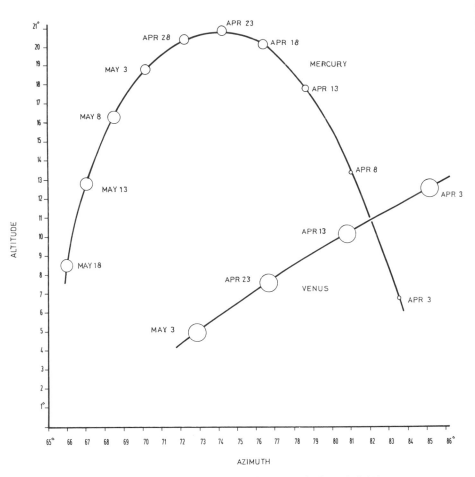

Figure 3. Morning apparition of Mercury for latitude S.35°.

though it will only be towards the end of the month before observers in the latitudes of the British Isles have any real chance of seeing the planet. Figure 4 also shows the path of Saturn amongst the stars throughout the year.

CLOSE APPROACH ASTEROIDS. Most of the asteroids or minor planets keep strictly to the region between the orbits of Mars and Jupiter, but there are some which depart from the main swarm, and may pass very close to the Earth. Careful searches, carried out largely by Mrs Eleanor Helin in America, have shown that there are many more of these 'close-approach' or Earth-grazing asteroids than has been believed. For many years the record was held by Hermes, discussed by Karl Reinmuth, which by-passed us on October 30, 1937 at less than twice the distance of the Moon – thereby causing some sensational newspaper headlines, such as 'Disaster Missed by Five Hours as Tiny Planet Hurtles Past!' But the record has since been broken, and on January 18, 1991 the asteroid known provisionally as 1991 BA passed by at only 106,000 miles (170,000 km), less than half the distance of the Moon. It has also the distinction of being the smallest asteroid ever tracked. Its diameter cannot be more than thirty feet.

Had it struck the Earth it would have caused widespread devastation, and there can be no doubt that collisions have often happened in the past. There is a theory – unproved, though backed by reasonably strong evidence – that about 65,000,000 years ago a strike by an asteroid altered the Earth's climate so dramatically that the dinosaurs could not cope with the changed conditions, and died out. Unquestionably there will be more collisions in the future; though the dangers are slight, it is fair to say that they are not nil.

One of the most famous 'Earth-grazers', Apollo (after which a whole class of Earth-crossing asteroids is named) was also discovered by Karl Reinmuth. It was subsequently lost for many years, but was recovered in 1973, and its orbit is now well known.

Reinmuth is one of this year's centenaries. He was born on April 4, 1892 at Heidelberg, and spent his entire career at the Königstuhl Observatory there. He concentrated upon asteroids, discovering 270 of them – including eight Trojans, which move in the same orbit as Jupiter. He left a collection of 12,500 asteroid positions as measured on the Königstuhl photographs – a tremendous labour, since in those days there were no computers and automatic measuring machines. He also found several comets. He died in 1979: Asteroid No. 1111 is named Reinmuthia in his honour.

GRIGORIJ SHAJN. By coincidence, our other main centenary honours another asteroid-hunter, though in this case his asteroid work was not his main research. Grigorij Abramovich Shajn was born on April 13, 1892 at Odessa; his father was a carpenter. Grigorij studied at home, and in 1910 published his first paper, dealing with the radiant of the Perseid meteors, in the journal of the Russian

Astronomical Society. Two years later he enrolled as a student at Tartu University in Estonia, but in 1914 he enlisted in the Russian Army, and served at the Front until 1917, after which he returned to his university studies. He joined the staff of the Pulkovo Observatory in 1921, and concentrated upon celestial mechanics; three years later he was put in charge of the new 40-inch reflector at the Simeis Observatory, and began work upon stellar spectroscopy. This was unceremoniously halted in 1941, when the Germans destroyed the telescope as well as the entire observatory. After the war the Observatory was rebuilt; Shajn became Director until 1952. During the last years of his life he was concerned mainly with the distribution of faint galactic nebulæ, and his *Atlas of Diffuse Gaseous Nebulæ*, published jointly with Miss V. E. Hase, contains many new objects. His one comet discovery was made much earlier, in 1925, while he was hunting for asteroids. He died in August 1956.

THATCHER'S COMET. The Lyrid shower, due this month, is not very rich, but it is well documented, and we are sure of its parent comet – 1861 I, discovered by A. E. Thatcher from New York. It became quite bright, with a maximum magnitude of 2.5, and a considerable tail. It was last seen on September 7, 1861, when it had declined to the tenth magnitude. The orbit was firmly established as elliptical, with a period of about 415 years, and there seems little doubt that it really is the parent comet of the Lyrid shower.

MAY

MERCURY, although not visible to observers in the latitudes of the British Isles because of the long twilight, continues to be visible as a morning object to observers further south for the first three weeks of the month. They should refer to Figure 3 given with the notes for April.

VENUS soon becomes lost in the morning twilight as it moves in towards superior conjunction next month.

MARS is a morning object, though still not visible to observers in the British Isles due to the lengthening twilight which has a greater influence on the visibility of the planet to these observers than the slowly increasing elongation from the Sun. It is quite surprising to note the difference in the dates for first detection of Mars as a morning object, as a function of latitude. Whereas observers at the Equator or further south could see it in mid-January, observers in the latitudes of the British Isles have to wait a further five months!

JUPITER, magnitude -2.1, is still visible as a brilliant object in the evening sky. By the end of the month it is not visible for long after midnight.

SATURN is visible as a morning object, magnitude $+0.7$. Figure 4 shows the path of Saturn amongst the stars throughout 1992.

Gamma Virginis. One of the most famous binary stars in the sky, Gamma Virginis, is well on view this month. It is very close to the celestial equator – in fact it is only just excluded from our list of 'equatorial stars', since its declination is S.01°27'. It has several proper names: Arich, Porrima, and Postvarta. The Chinese called it Shang Seang, the High Minister of State.

Both the components are of magnitude 3.5, giving a combined naked-eye magnitude for the star of 2.75, and both are of spectral type F0 – so that they are in theory slightly yellowish, though

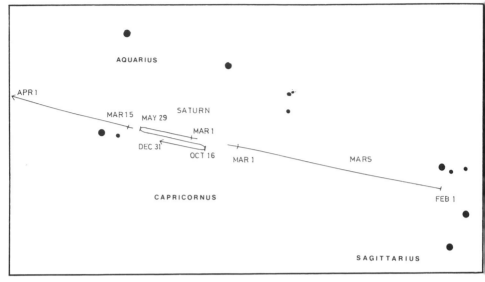

Figure 4. The paths of Mars and Saturn.

almost all observers will certainly call them white. The distance is only 36 light-years, and each component is about 8 times as luminous as the Sun.

A few decades ago Gamma Virginis was one of the easiest telescopic binaries in the sky. Its double nature was discovered in 1718 by Bradley and Pound; Sir John Herschel calculated the orbit, and predicted that by 1836 the components would be so close together that only the world's greatest telescopes would separate them. This was correct. Subsequently the separation increased, reaching 6″.2 in 1920. The period is 171.4 years, and the separation is now becoming much less; by the end of the century it will be a difficult double, and at the time of closest approach, in 2007, the star will appear single in all but very large instruments. This does not, of course, indicate that there is any danger of collision! The orbit is decidedly eccentric, and the real separation ranges between 3 astronomical units and 70 astronomical units. (One astronomical unit, as we have seen, is the mean distance between the Earth and the Sun: 93 million miles or 150 million kilometres.)

A GREAT PIONEER PHOTOGRAPHER. Around a century ago photography replaced the human eye as the main technique in most branches of astronomy; only now is photography itself being superseded by electronic devices. It is often forgotten that the very word 'photography' was coined by an astronomer – Sir John Herschel, son of the discoverer of Uranus.

This month's centenary honours one of the greatest of the early astronomical photographers, Lewis Morris Rutherfurd (it is remarkable how often his name is mis-spelled 'Rutherford'!). He was born on November 25, 1816, and trained as a barrister, but after he built an observatory at his home in New York he devoted his time to astronomy and abandoned his legal career. With the first lens ever made specifically for celestial photography he took fine photographs of the Moon during the 1850s, and he also took excellent photographs of the Sun; his ruled gratings for solar spectra were the best made before those of Rowland. Rutherfurd took splendid pictures of star clusters, and demonstrated the advantages of measuring star positions photographically. He also became a pioneer of stellar spectroscopy. He died a hundred years ago, on May 20, 1892, by which time the power of photographic methods in astronomy had become clear to all.

CURIOUS SUGGESTIONS. Now and then we come across a suggestion which seems to be a little strange, to put it mildly. One of these came from the great Russian astrophysicist Iosif Shklovskii in 1959, when he suggested that the two dwarf satellites of Mars, Phobos and Deimos, might be artificial space-stations. It was only many years later, in an article which I invited him to write for the *Yearbook*, that he retracted this and said that it had never been more than a practical joke.

However, another idea concerning the Moon seems to have been quite serious – and since the Moon is full in the middle of this month, we may pause to say something about it. Professor Alexander Abian, of Iowa University in the United States, seems to have come to the conclusion that we could alter the Earth's orbit, and improve our climate, by either blowing up the Moon or else disrupting it and dumping parts of it in the Pacific Ocean. He and his colleagues petitioned Congress to this effect. It is hardly surprising that astronomers were unimpressed, and that when asked to comment many of them (including your present Editor) had difficulty in keeping a straight face. Abian seems to have outshone even the renowned Immanuel Velikovsky, who believed that Venus was formerly a comet which bounced around the Solar System rather in the manner of a cosmic table-tennis ball!

JUNE

New Moon: June 1, June 30   *Full Moon:* June 15

Solstice: June 21

MERCURY is unsuitably placed for observation from northern temperate latitudes. However, after the first ten days of the month it is visible as an evening object to observers nearer the Equator and in southern latitudes. It will then be seen low in the west-north-western sky about the time of ending of evening civil twilight.

VENUS passes through superior conjunction on June 13 and is therefore too close to the Sun for observation. Indeed for a short while on that date it could safely be said that no Earth-based instrument could detect it since it is actually occulted by the Sun.

MARS, magnitude +0.9, is a morning object. Observers in the British Isles may expect their first sight of the planet during the month, low above the eastern horizon about two hours before sunrise.

JUPITER is still a brilliant object in the western sky in the evenings.

SATURN, magnitude +0.6, continues to be visible as a morning object.

This Month's Total Solar Eclipse. Total eclipses often seem to occur in inaccessible places. Last year's event, when the track crossed Mexico and Hawaii, was an exception. The present track is by no means unfavourable, and the maximum length of totality is 5 minutes 26 seconds – so that many eclipse parties will be hoping for good weather on June 30.

The total solar eclipses for the rest of the century are as follows:

Date	Max. length of totality	Area
1994 Nov. 3	4m 23s	Peru, Brazil, South Atlantic
1995 Oct. 24	2m 5s	Iran, India, East Indies, Pacific
1997 Mar. 9	2m 50s	Russia, Arctic
1998 Feb. 26	3m 56s	Pacific, Atlantic
1999 Aug. 11	2m 23s	Atlantic, England, France, Turkey, India

No doubt plans are already being made for the 1999 eclipse, when the track will cross Cornwall. Hotel bookings have already been made!

BETA LIBRÆ. Libra, the Scales or Balance, is one of the least impressive constellations of the Zodiac; indeed, it was once called Chelæ Scorpii (the Scorpion's Claws), and one of its leading stars, Sigma Libræ, has been stolen from Scorpius – it used to be known as Gamma Scorpii. Libra, well seen this month, has no really distinctive shape, and there is little here of immediate interest, though Delta Libræ is one of the brightest of the Algol-type eclipsing binaries (the magnitude range is from 4.9 to 5.9, in a period of 2.33 days).

Perhaps the most notable star in the group is Beta Libræ, which has the somewhat barbarous proper name of Zubenelchemale, the 'Northern Claw'. It is about 120 light-years away, and around 150 times as luminous as the Sun, with a spectral type of B8. The apparent magnitude is 2.61.

There are two points of interest concerning it. One is the fact that it has been suspected of fading in recorded times. According to one account (which admittedly is probably not reliable) the Greek philosopher Eratosthenes classed it as of the first magnitude, and equal to Antares. Ptolemy, around AD 150, also compared it with Antares, but since he gave both stars as of the second magnitude rather than the first this is not significant. Before 1700 the magnitude was generally given as 2, whereas it is now slightly below 2.5; but it would be dangerous to place much reliance on old estimates, and the possibility of real, permanent change in Beta Libræ seems rather remote.

More interesting are the persistent reports that it has a greenish hue. The only stars which appear decidedly green to visual observers are the companions of red primaries, and the greenness is then due to contrast. But Beta Libræ is single. The Rev. T. W. Webb, author of the classic book *Celestial Objects*, referred to its 'beautiful pale green hue, very unusual among conspicuous stars', while the American writer on astronomy W. T. Olcott said that it was 'the only naked-eye star that is green in colour'.

Whether or not there is any perceptible green colour is by no means certain; most observers will certainly call it white. However, the suggestion has been made so often that the star is worth examination.

The other leader of the constellation, Zubenelgenubi or Alpha Libræ (the 'Southern Claw') is of magnitude 2.7; it has a companion of magnitude 5.2, separation 231 seconds of arc, so that this is a very easy binocular pair. Since the two share a common motion through space they are presumably associated, but they must be at least 5000 astronomical units apart.

AN ASTRONOMICAL ADMIRAL. Our centenary this month honours Ernest Amédée Barthélémy Mouchez, who was born in Madrid, of French parentage, in 1821 and died in Paris on June 29, 1892. He joined the French Navy in 1839, and was a captain by 1861; he observed a transit of Venus from the island of St Paul, and carried out coastal surveys of Africa and South America. In 1878, on the death of Le Verrier, he became Director of the Paris Observatory, and given the rank of rear-admiral. He did much to modernize the Observatory, and had a large share in the organization of the International Astrographic Chart, presiding at the inaugural meeting of its Committee. A lunar crater is named after him.

JULY

New Moon: July 29   *Full Moon:* July 14

EARTH is at aphelion (farthest from the Sun) on July 3 at a distance of 152 million kilometres.

MERCURY cannot be seen from the latitudes of the British Isles because of the long duration of twilight. From latitudes further south it continues to be visible as an evening object, except for the last week of the month. Greatest eastern elongation occurs on July 6.

VENUS is unsuitably placed for observation being so close to the Sun, though at the very end of the month observers in the tropics may glimpse it low above the west-north-western horizon for a short while after sunset.

MARS continues to be visible as a morning object, magnitude +0.8. At the very end of the month Mars will be seen moving eastwards between the Pleiades and the Hyades.

JUPITER, magnitude −1.8, continues to be visible as a brilliant object in the evening sky. By the end of the month it will be rather difficult for observers in the British Isles to detect because of the long duration of twilight.

SATURN continues to be visible as a morning object, magnitude +0.4.

URANUS is at opposition on July 7, in Sagittarius. The planet is only just visible to the naked eye under the best of conditions (magnitude +5.6) and in a small telescope it appears as a slightly greenish disk. At opposition Uranus is 2771 million kilometres from the Earth.

Uranus is slowly catching Neptune up and in heliocentric longitude the separation is only 1.5 degrees. Shortly before opposition in 1993 it will have overtaken Neptune.

NEPTUNE is at opposition on July 9, also in Sagittarius. It is not visible to the naked eye since its magnitude is +7.9. The distance of Neptune from the Earth at opposition is 4364 million kilometres.

THE TELESCOPIC PLANETS. The two outer giants, Uranus and Neptune, come to opposition during the same week in July. Both are in Sagittarius, so that they are unfavourably placed for British observers because they are so low down. Uranus, at magnitude 5.6, is distinctly visible with the naked eye, though it looks exactly like a dim star, and it is hardly surprising that its planetary nature was not appreciated until 1781, when it was identified by William Herschel. Neptune is well below naked-eye visibility, though binoculars will show it.

The only asteroid ever visible with the naked eye is Vesta, which can reach magnitude 5.6 when at its best. The four asteroids Ceres, Pallas, Juno, and Vesta were discovered between 1801 and 1809, after which there was a delay until 1845, when the German amateur Hencke found asteroid No. 5, Astraea. Astraea, with a mean opposition magnitude of 9.8, is much fainter than Neptune, but of course it is much closer to the Sun, so that its quicker movement against the starry background makes it easier to detect.

All the same, it does sometimes seem a little surprising that the great observers such as Herschel failed to locate any more of the bright telescopic or near-telescopic members of the Solar System. Those with mean opposition magnitudes above 9 are as follows:

Uranus	5.7	Juno	8.1
Vesta	6.4	Hebe	8.3
Ceres	7.3	Eunomia	8.5
Pallas	7.5	Flora	8.7
Neptune	7.7	Melpomene	8.9
Iris	7.8		

However, only the two giant planets show measurable disks.

It is interesting to recall that during Galileo's studies of the four major satellites of Jupiter, in 1610, he recorded a 'star' which was unquestionably Neptune! Therefore, there was a possibility that

Neptune could have been identified long before it was finally tracked down in 1846.

THE SOUTHERN TRIANGLE. Very few of the constellations bear the slightest resemblance to the objects they are supposed to represent. It takes a lively effort of the imagination to make a bull out of Taurus, for example, and most of the 'modern' constellations are even worse; how does one twist the dim stars of Microscopium into a figure of a microscope? But there are a few exceptions, and one of these is Triangulum Australe, the Southern Triangle.

It lies in the far south, and is therefore never visible from European latitudes. There are three stars above the third magnitude, and they really do make a well-formed triangle:

	R.A.			Declination			Magnitude	Spectrum
	h	m	s	°	′	″		
Alpha	16	48	39.8	−69	01	39	1.9	K2
Beta	15	55	08.4	−63	25	50	2.8	F5
Gamma	15	18	54.5	−68	40	46	2.9	A0

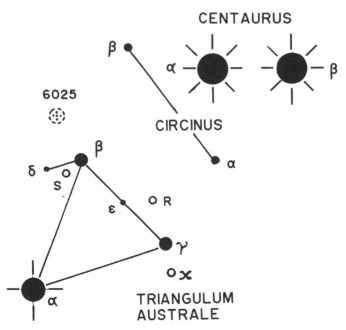

The other stars above the fifth magnitude are Delta (3.9), Epsilon (4.1), and Zeta (4.9), but these do not destroy the main pattern; Epsilon lies midway between Beta and Gamma, while Delta is close to Beta.

There is no difficulty in locating the Southern Triangle, because it lies close to Alpha and Beta Centauri, the brilliant Pointers to the Southern Cross. Moreover, Alpha Trianguli Australe, with its K2-type spectrum, is distinctly orange. There is not much else of real note in the constellation, apart from the open cluster NGC 6025, which has an integrated magnitude of about 5 and contains at least sixty stars; R Trianguli Australe is a Cepheid with a range of from magnitude 6.4 to 6.9 and a period of 3.4 days, while S Trianguli Australe is also a Cepheid, varying from 6.1 to 6.8 in a period of 6.3 days. The little constellation of Circinus, the Compasses, lies between Triangulum Australe and the Pointers.

AUGUST

New Moon: August 28   *Full Moon:* August 13

MERCURY passes through inferior conjunction on August 2 but becomes visible as a morning object after the first ten days of August. By the end of the month it is two magnitudes brighter than at the middle of the month, changing from +1.0 to −1.1.

VENUS is moving slowly outwards from the Sun and is visible as an evening object, magnitude −3.9. It is still a difficult object for observers in northern temperate latitudes who will only see it for a very short while after sunset, low above the western horizon.

MARS, magnitude +0.7, is a morning object and well placed for observation during the second part of the night in the eastern and south-eastern skies. Early in the month it passes north of Aldebaran. Both objects have almost exactly the same brightness and both have a reddish tinge. Figure 5 shows the path of Mars amongst the stars for the rest of the year.

JUPITER, magnitude −1.7, is moving steadily in towards the Sun and can only be seen for a short while in the evenings in the western sky. For observers in the British Isles Jupiter is disappearing into the evening twilight and will not become visible again until October.

SATURN, magnitude +0.2, is at opposition on August 7 and therefore visible from dusk to dawn. At opposition Saturn is 1329 million kilometres from the Earth.

THE 1933 WHITE SPOT ON SATURN. The 1990 white spot on Saturn was spectacular; many people will have seen drawings and photographs of it – notably the splendid image obtained from the Hubble Space Telescope. Before long it lengthened until it had become a white zone rather than a well-defined spot. This also happened with

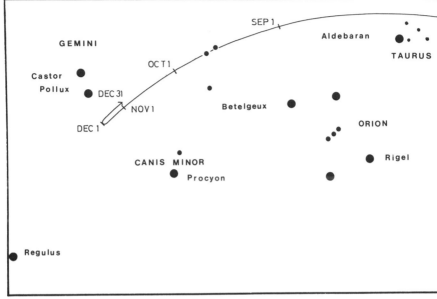

Figure 5. The path of Mars.

the great spot of 1933, and it is worth looking back at this remarkable event.

The spot was sighted on August 3, 1933 by W. T. Hay, using a 6-inch refractor at his observatory in outer London. (Hay's fame as a stage and screen comedian, as 'Will Hay', is so great that one tends to forget that he was also an expert amateur astronomer.) Hay detected a large white patch in Saturn's equatorial zone; it was oval, and about one-fifth of the planet's diameter in length, both ends being well-defined. Hay at once telephoned the famous amateur astronomer Dr W. H. Steavenson, who confirmed the spot a few minutes later. It was observed at about the same time, independently, by A. Weber from Berlin, but Hay was the first to announce the discovery.

On the days after the discovery the spot lengthened quickly in the preceding direction, and by August 6 it occupied the whole width of the zone in latitude, even at one time encroaching on the north equatorial belt. On August 9 Hay found that it took fifty-one minutes to cross the planet's central meridian, and he wrote that

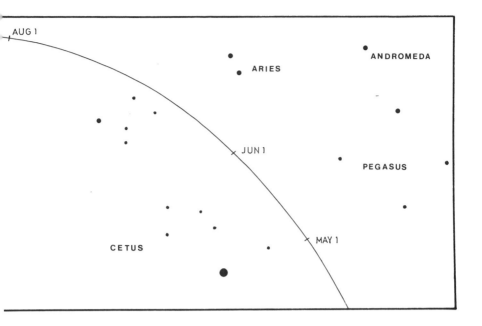

'following on some disturbance beneath the bright matter which forms the equatorial zone, there was a sudden rush or concentration of some of this bright matter towards a centre, uncovering at each end of the concentration, the dark matter which possibly lies at a lower altitude'.

By mid-September the spot was no longer a 'spot', but merely a brighter part of the equatorial zone. It seems to have appeared suddenly, because Saturn had been examined from the Lowell Observatory in Arizona only forty-eight hours earlier, and there had been no sign of anything unusual. Steavenson commented that at its best it was elliptical, and about the same proportions as the Great Red Spot on Jupiter; the mean rotation period was found to be about 10 hours 14 minutes.

An earlier white spot had been recorded in 1876 by Hall; it was found on December 7, and followed for some weeks before it too lost its identity and became a bright zone. The derived rotation period was 10 hours 14½ minutes, in close agreement with that of the spot of 1933.

One very interesting point is that the intervals between the three white spots – 1876, 1933, and 1990 – are equal at fifty-seven years. The underlying cause of this is by no means clear, and of course it may be nothing more than coincidence, but one imagines that Saturn observers will be very much on the alert at the opposition of 2047!

THE 1992 PERSEIDS. The Perseid meteor shower is normally much the richest and most reliable of the year. Unfortunately conditions in 1992 are almost as bad as they can be, because the maximum of the shower almost coincides with the time of full moon. On the other hand, it would be wrong to suggest that nothing useful can be done by visual observers, and it will be particularly worth while to keep watch for really brilliant meteors which can show up even against the moonlit sky.

R CORONÆ BOREALIS. The little constellation of the Northern Crown is easy to locate; it adjoins Boötes (the Herdsman), and is made up of a semi-circle of stars, of which the brightest, Alphekka, or Alpha Coronæ, is of the second magnitude. Inside the 'bowl' lies one of the most famous variable stars in the sky, R Coronæ. Usually it is on the fringe of naked-eye visibility, but at irregular intervals it fades down so considerably that it cannot be seen at all except with large telescopes; it takes some time to recover. Apparently it hides itself behind clouds of soot in its atmosphere.

If you look at Corona with binoculars, you should see two stars above magnitude 7. If only one is on view, you may be sure that R Coronæ is undergoing a minimum. As these words are being written in March 1991, it is quite impossible to say whether or not R Coronæ will be faint by the time that this *Yearbook* appears in print!

SEPTEMBER

New Moon: September 26   *Full Moon:* September 12

Equinox: September 22

MERCURY is virtually unobservable during September, though observers in the tropics may be able to see it as a difficult morning object for the first two or three days of the month, low above the eastern horizon before dawn.

VENUS, magnitude −3.9, continues to be visible in the evenings shortly after sunset, low above the western horizon. Northern Hemisphere observers will still only experience a relatively short period of observation after sunset since the motion of the planet away from the Sun is counteracted by its rapid southward movement in declination.

MARS, magnitude +0.4, is still a morning object, moving from Taurus into Gemini during the month.

JUPITER remains too close to the Sun for observation as it passes through conjunction on September 17.

SATURN, magnitude +0.4, is just past opposition and thus visible for the greater part of the night.

THE FIRST MOON LANDING. When asked the date of the first Moon landing, most informed people would probably say '1969', when first Neil Armstrong and then Edwin Aldrin stepped out from the lunar module *Eagle* on to the rocks of the Mare Tranquillitatis. Yet this is not strictly true. Armstrong and Aldrin were the first men to land, but the first impact of a man-made object on to the Moon occurred much earlier on September 13, 1959, twenty-three years ago now.

The vehicle was Luna 2, the second of the Russian space-craft of that year – Luna 1 had already by-passed the Moon, in January 1959, and had confirmed that there is no detectable overall lunar magnetic field. Luna 2 was different. It began its journey on September 12, and, like its predecessor, it sent back information of various kinds during its flight; in particular it transmitted data about cosmic rays and the numbers of tiny meteoritic particles in space. The lack of any lunar magnetic field was again confirmed.

Before hitting the Moon, the last stage of the compound rocket was separated from the container carrying the scientific instruments. It seems that no attempt was made to bring the capsule down gently (and this would indeed have been a very difficult matter in 1959, when the Space Age was little more than two years old). The predicted time of impact was 21 hours 0 minutes on September 13. Visual observers were very much on the alert, but, more importantly, so were the radio astronomers at Jodrell Bank, where the 250-foot radio telescope – now called the Lovell Telescope – was in operation. The Jodrell Bank data showed that as the probe neared its target it was still sending back loud, clear signals. At 21 hours 2 minutes 23 seconds the signals ceased abruptly. This, then, was the moment of impact.

The precise landing site is uncertain, but is probably near 30°N. 1°W., not far from the great walled plain Archimedes in the Mare Imbrium (Sea of Showers). We will find out for certain only when an expedition examines the area and locates the shattered remains of Luna 2.

No data concerning the Moon's surface were sent back, but all the same Luna 2 has an important place in history: it was the first vehicle ever to be sent direct from one world to another. And less than ten years elapsed between the crash-landing of Luna 2 and Neil Armstrong's 'one small step'.

LIGHT POLLUTION. Much has been heard lately of 'light pollution'; the skies are no longer dark at night, and anyone living in or near a city has little chance of seeing any but the brightest stars. Since September is the month when the Sun reaches the celestial equator, to spend the next half-year in the Southern Hemisphere, this may be the moment to clear up one popular misconception about what astronomers want.

The claim is that astronomers are anxious to 'put the lights out', thereby increasing the risk of crime in darkened streets. This is completely wrong, and no serious astronomer has ever suggested anything of the sort. The point is that present lighting systems are wasteful and damaging, because so much of their illumination is directed upward instead of down – which not only pollutes the sky, but also wastes a great deal of money.

Therefore, say the astronomers, what has to be done is to make sure that any new installations are efficient in this respect. A few cities have already started to take notice, and there was an interesting and welcome development in Cumbria, in northern England, in early 1991, when after representations from local astronomers the parish council of Great Salkend, which lies in the Eden Valley between Penrith and Carlisle, passed a resolution which required that all new and replacement street lighting in the village should direct light *towards* the ground.

Let us hope that other authorities follow the example of Great Salkend. It would indeed be sad to turn whole countries into areas from which the stars can never be properly seen.

OCTOBER

New Moon: October 25   *Full Moon:* October 11

Summer Time in Great Britain and Northern Ireland ends on October 25.

MERCURY is not suitably placed for observation for observers in the British Isles, though it is an evening object for observers further south. For observers in southern latitudes this will be the most favourable evening apparition of the year. Figure 6 shows, for observers in latitude S.35°, the changes in azimuth (true bearing from the north through east, south, and west) and altitude of Mercury on successive evenings when the Sun is 6° below the horizon. This condition is known as the end of evening civil twilight, and in this latitude and at this time of year occurs about 30 minutes after sunset. The changes in the brightness of the planet are indicated by the relative sizes of the circles marking Mercury's position at five-day intervals. It will be noticed that Mercury is at its brightest before it reaches greatest eastern elongation (24°) on October 31. Its magnitude remains near 0 throughout October.

VENUS, magnitude −4.0, continues to be visible as an evening object, low in the south-western sky after sunset. Although 36° from the Sun by the end of the month it is only visible to observers in the British Isles for about half-an-hour after sunset.

MARS, magnitude 0.0, is a bright object in Gemini, now visible before midnight.

JUPITER, after the first week of the month, emerges from the morning twilight and becomes visible as a brilliant morning object, magnitude −1.7. It will be seen low above the eastern horizon before dawn.

SATURN continues to be visible as an evening object, magnitude +0.5. Although still well placed for observation by those in the Southern Hemisphere, observers in the latitudes of the British Isles will only see it low in the south and south-western skies.

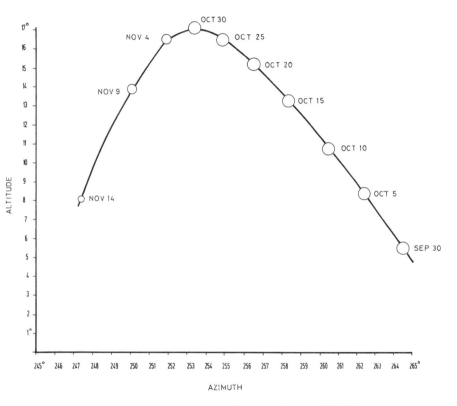

Figure 6. Evening apparition of Mercury for latitude S.35°.

THE TWO SPIRALS. Two of the most famous of all spiral galaxies are on view this month: Messier 31 in Andromeda, and Messier 33 in Triangulum.

Everyone with even a passing interest in astronomy knows about M.31, the Andromeda Spiral (once known as the Andromeda Nebula, before it was realized that nebulæ and galaxies are very different classes of objects). It is easily visible with the naked cye,

107

and in AD 964 the Arab astronomer Al-Sufi called it 'a little cloud' –
though, amazingly, the great Danish observer Tycho Brahe said
nothing about it. It was first seen through a telescope on December
15, 1612 by Simon Marius, who said that it looked like 'the flame of
a candle seen through horn', or like 'a cloud consisting of three rays;
whitish, irregular and faint, brighter toward the centre'. To Ed-
mond Halley it seemed to send out 'a radiant beam', and on August

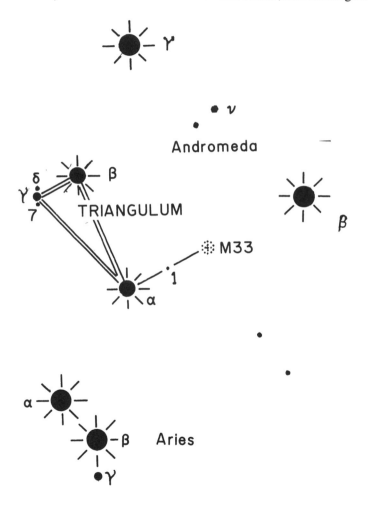

3, 1764 Charles Messier himself described it as a 'beautiful nebula, shaped like a spindle . . . No star recognized. Resembles two cones or pyramids . . . diameter 40 minutes of arc.' The spiral structure was clearly shown on a photograph taken by Roberts in 1888, while the Earl of Rosse, using his 72-inch reflector at Birr Castle from 1845, considered – rightly – that it could be resolved into stars.

M.31 is placed at a narrow angle to us, so that the full beauty of the spiral is lost, but photographs show it well; visually, with a small telescope, it is admittedly rather a disappointment. It lies at a distance of 2,200,000 light-years, and is a system decidedly larger than our own Galaxy.

M.31 contains objects of all kinds. Many novæ have been seen in it, and one supernova – S Andromedæ of 1885, which reached the fringe of naked-eye visibility even though its true nature was not appreciated at the time. It is only in modern years that its remnant has been re-located.

The other spiral, M.33 in Triangulum, is much less obvious. It was discovered by Messier in 1764, and described as 'A whitish light of almost even brightness. However, along two-thirds of its diameter it is a little brighter. Contains no star.' To William Herschel it had 'a mottled aspect'; to Admiral Smyth it was 'a large, distinct but faint and ill-defined, pale white nebula'.

There seems little doubt that keen-eyed people can just glimpse M.33 without optical aid, but it is extremely difficult, and those observers with no more than normal sight will probably have to admit failure. With binoculars it is easy to locate. Using a magnification of ×7, it is in the same field with Alpha Trianguli, past the fainter star 1 Trianguli and at about an equal distance from it in the direction of Beta Andromedæ (Mirach). With a magnification of ×12 it is still just in the same field as Alpha Trianguli.

Strangely enough, M.33 is sometimes very elusive telescopically, because of its low surface brightness; it is certainly easier to locate with binoculars than with a small telescope which is not specially designed for rich-field work. The spiral is much more open than with M.31, because it is placed at a more favourable angle to us, but there is no chance of seeing the spiral arms with a small telescope. M.33 is a much smaller galaxy than M.31, and is slightly further away, though it is, of course, a member of the Local Group.

NOVEMBER

New Moon: November 24   *Full Moon:* November 10

MERCURY continues to be visible as an evening object for the first half of the month, for observers in tropical and southern latitudes. After passing through inferior conjunction on November 21 it moves rapidly away from the Sun to become visible as a morning object, magnitude around +0.5, for the last few days of the month. Under good conditions it may then be detected low above the south-eastern horizon around the time of beginning of morning civil twilight.

VENUS, magnitude −4.1, is a brilliant object, visible above the south-western horizon after sunset.

MARS is now brightening noticeably, its magnitude increasing steadily during the month from −0.2 to −0.8. By the end of November it is on the border of Gemini and Cancer, in almost a direct line with Castor and Pollux. This position marks its first stationary point.

JUPITER, magnitude −1.7, is a brilliant morning object, moving outwards from the Sun and steadily becoming visible for longer and longer each morning.

SATURN, magnitude +0.7, is still visible as an evening object, in Capricornus.

THE BIELID METEORS. The most famous meteor shower of November is that of the Leonids, which are generally rather sparse, but which may at intervals produce really magnificent displays – as in 1799, 1833, 1866, and 1966. We may reasonably hope for another Leonid 'storm' in 1998 or 1999. But there has also been another fascinating November shower, that of the Bielids or Andromedids, even though it now seems to have virtually died out.

The parent comet was originally seen by the French astronomer Montaigne in 1772; it was recovered by Pons in 1805, and then by von Biela in 1826. It was found to have a period of 6.75 years, and is always known as Biela's Comet. It was missed in 1839 because it was badly placed, but returned in 1845, and in 1846 caused an astronomical sensation by dividing into two. The twins came back on schedule in 1852, were missed in 1859 because they were again badly placed, and were confidently expected once more in 1865–6. To the general surprise, they failed to appear, and in fact they have never been seen since, but in 1872 a brilliant meteor shower was seen on November 27, quite clearly and unmistakably associated with the débris of Biela's Comet. Over a period of six hours, an Italian team headed by P. F. Denza recorded over 33,400 meteors.

The shower was not repeated in the following years, but on November 27, 1885 the Andromedids were back, and the famous meteor observer W. F. Denning, at Bristol, recorded that for a while 'meteors were falling so thickly as the night advanced that it became almost impossible to enumerate them'.

This was the last really major Andromedid display, but the shower lingered on at intervals; on November 24, 1899 and November 21, 1904 the ZHR was between 100 and 200 for a while. (The ZHR or Zenithal Hourly Rate is the number of naked-eye meteors of a shower which would be expected to be seen by an observer under ideal conditions, with the radiant at the zenith. In practice, of course, these ideal conditions are virtually never achieved, so that the actual ZHR is always rather less than the theoretical value.) There was a minor Andromedid shower in 1940, and vague indications during the 1950s, but it seems that the displays are now more or less over, though there are indications of a few Andromedids around November 14 each year; the ZHR is never more than 5 – very different from the spectacular displays after the original disappearance of the comet.

Of course, it is always dangerous to be over-confident. There have been comets which have been lost for many years, only to be recovered much later; Holmes' Comet of 1892 is a classic case – it was a naked-eye object in that year, but after 1906 it was seen no more until the return of 1965, when it was picked up as a very faint object and has been followed at most returns since. So there are still occasional searches for Biela's Comet – and, incidentally, it was during one of these searches, in 1973, that Dr Lubos Kohoutek from Hamburg discovered the comet which was expected to become brilliant, but which signally failed to do so. But all in all, it seems that with the demise of the Andromedids we really must say good-bye to Biela's Comet in any form.

THE W OF CASSIOPEIA. Cassiopeia, the proud queen of the Perseus legend, is one of the most famous of all the constellations. It lies in the far north; from Britain it is near the zenith during November evenings, and it cannot be mistaken. It and the Great Bear, Ursa Major, lie on opposite sides of the Pole Star and about equally distant from it. Neither Ursa Major nor Cassiopeia ever sets from Britain.

Of the five stars in the familiar W or M shape, Epsilon is of magnitude 3.4 and Delta of magnitude 2.7, while Beta is brighter at 2.3. The central star of the W, Gamma Cassiopeiæ, is variable. It is an unstable star, and its behaviour is unpredictable; it has been known to rise to magnitude 1.6, but it has also been known to become as faint as 3.3. Alpha Cassiopeiæ or Shedir, an orange star of type K, is officially listed as of magnitude 2.2, but there are strong suggestions that it may be variable over a small range – perhaps 2.1 to 2.5. So it is always worth looking at the W and trying to decide which is its brightest member; for over ten years now there has been very little difference between Gamma, Beta, and Alpha.

DECEMBER

New Moon: December 24   *Full Moon:* December 9

Solstice: December 21

MERCURY is a morning object, though for observers in the latitude of the British Isles the period of visibility is restricted to the first three weeks of the month. For observers in northern temperate latitudes this will be the most favourable morning apparition of the year. Figure 7 shows, for observers in latitude N.52°, the changes in azimuth (true bearing from the north through east, south and west) and altitude of Mercury on successive evenings when the Sun is 6° below the horizon. This condition is known as the beginning of morning civil twilight, and in this latitude and at this time of year occurs about 35 minutes before sunrise. The changes in the brightness of the planet are indicated by the relative sizes of the circles marking Mercury's position at five-day intervals. It will be noticed that Mercury is at its brightest after it reaches greatest western elongation (21°) on December 9. Its magnitude range is fairly small, the actual magnitude being +0.5 at the beginning of the month and −0.5 at the end.

VENUS, magnitude −4.2, continues to be visible as a magnificent object in the south-western sky in the evenings. As it moves northwards in declination observers in the British Isles obtain a longer period of observation – by the end of the year it extends to over three hours after sunset.

MARS is now visible as a prominent object in the night sky and visible for the greater part of the night, though it is another three months before it comes to opposition. During December its magnitude brightens from −0.8 to −1.3. By the end of the month observers in the British Isles will be able to see it low in the east-north-east as soon as it gets dark.

JUPITER continues to be visible as a brilliant morning object, in the south-eastern sky, magnitude −1.9. Jupiter is in Virgo.

SATURN continues to be visible as an evening object in the south-western sky. Its magnitude +0.7.

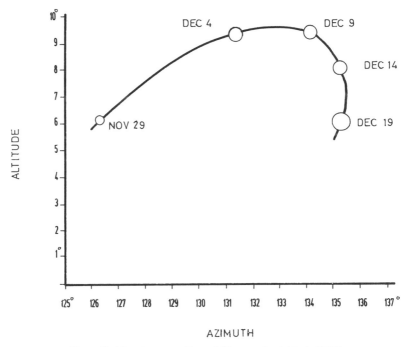

Figure 7. Morning apparition of Mercury for latitude N.52°.

THIS MONTH'S LUNAR ECLIPSE. The eclipse of December 9–10 will be among the most favourable of the present decade, since totality will last for over an hour, and the eclipse will be seen from Europe – including Britain – as well as parts of Africa and North America.

The colour of the eclipsed Moon depends wholly upon conditions in the Earth's atmosphere, through which all sunlight reaching the Moon must pass. There can be great differences between different eclipses. That of 1848 is said to have been so 'bright' that many observers refused to believe that it was happening at all, while in

1816 the Moon became invisible with the naked eye. The French astronomer A. Danjon has attempted to draw up an eclipse scale from 0 (very dark) to 4 (bright) and to correlate it with sunspot activity, though the evidence is still very meagre. During a lunar eclipse, a wave of cold sweeps across the Moon's surface, and in the past it has been claimed that there have been marked effects upon certain features. Of special note in this respect is Linné, on the Mare Serenitatis (Sea of Serenity), which was recorded as a small but deep crater before 1866 and as a pit surrounded by a white nimbus afterwards. Many suggestions have been made that a change occurred there, but this now seems highly unlikely, particularly as space-craft photographs show Linné as a perfectly normal small crater. W. H. Pickering, who paid great attention to the Moon around 1920, believed that the sudden drop in temperature caused 'snow' to be deposited around Linné, and that this made the white nimbus appear more conspicuous after eclipse than before; but this again seems improbable in the highest degree.

On the other hand, we do know that different parts of the Moon have different surface texture, and it is important to make careful surveys. Something important may be found.

The total lunar eclipses for the rest of the century are as follows:

Date	Time of mid-eclipse, GMT		Length of totality	
	h	m	h	m
1993 June 4	13	02	1	36
1993 Nov. 29	06	26	0	56
1996 Apr. 4	00	11	1	26
1996 Sept. 27	02	55	1	10
1997 Sept. 16	18	47	1	2
2000 Jan. 21	04	45	1	16
2000 July 16	13	57	1	0

Note that the two eclipses of 2000 have been included in 'the present century' – and this question is bound to crop up time and time again as the year 2000 is approached. Yet there can be no argument about it. The present century began on January 1, 1901, and will end on December 31, 2000. So the first day of the twenty-first century will be January 1, 2001, and not January 1, 2000.

It does not now seem very long to wait!

Eclipses in 1992

In 1992 there will be five eclipses, three of the Sun and two of the Moon.

1. *An annular eclipse of the Sun on January 4–5* is visible as a partial eclipse from the Oceania, Philippines, Japan, extreme coast of north-east Asia, northern part of Australia, Pacific Ocean and west coast of North America. The eclipse begins on January 4 at $20^h 04^m$ and ends on January 5 at $02^h 06^m$. The annular phase begins on January 4 at $21^h 16^m$ in the Pacific Ocean and ends on January 5 at $00^h 53^m$ on the coast of North America. The maximum duration of the annular phase is $11^m 36^s$.

2. *A partial eclipse of the Moon on June 15* is visible from Antarctica, Africa except the eastern part, southern tip of Greenland, South America, North America except the north-west, Central America, part of the Pacific Ocean and New Zealand. The eclipse begins at $03^h 27^m$ and ends at $06^h 27^m$. The time of maximum eclipse is $04^h 57^m$ when 0.68 of the Moon's diameter is obscured.

3. *A total eclipse of the Sun on June 30.* The path of totality begins on the coast of Uruguay, crosses the South Atlantic and ends in the Southern Ocean south of Africa. The partial phase is visible from central South America, the South Atlantic, south-west of Africa, Southern Ocean south of Madagascar and the extreme south-west Indian Ocean. The eclipse begins at $09^h 51^m$ and ends at $14^h 30^m$; the total phase begins at $11^h 02^m$ and ends at $13^h 19^m$. The maximum duration of totality is $5^m 26^s$.

4. *A total eclipse of the Moon on December 9–10* is visible from Asia except the extreme east, Europe including the British Isles, Africa, Atlantic Ocean, Iceland, Greenland, South America except the south, Central America, North America except western coast. The eclipse begins on December 9 at $22^h 00^m$ and ends on December 10 at $01^h 28^m$. Totality lasts from $23^h 07^m$ on December 9 to $00^h 21^m$ on December 10.

5. *A partial eclipse of the Sun on December 23–24* is visible from eastern China, Korea, Japan, the extreme east of USSR and south-west Alaska. The eclipse begins on December 23 at $22^h 21^m$ and ends on December 24 at $02^h 41^m$. At time of maximum eclipse 0.84 of the Sun's diameter is obscured.

Occultations in 1992

In the course of its journey round the sky each month, the Moon passes in front of all the stars in its path and the timing of these occultations is useful in fixing the position and motion of the Moon. The Moon's orbit is tilted at more than five degrees to the ecliptic, but it is not fixed in space. It twists steadily westwards at a rate of about twenty degrees a year, a complete revolution taking 18.6 years, during which time all the stars that lie within about six and a half degrees of the ecliptic will be occulted. The occultations of any one star continue month after month until the Moon's path has twisted away from the star but only a few of these occultations will be visible at any one place in hours of darkness.

There are six occultations of bright planets in 1992, two of Mercury, two of Venus, and two of Mars. One occultation of Mercury is visible from Europe, the other from North America.

Only four first-magnitude stars are near enough to the ecliptic to be occulted by the Moon; these are Regulus, Aldebaran, Spica, and Antares. No occultations of these stars occur in 1992.

Predictions of these occultations are made on a world-wide basis for all stars down to magnitude 7.5, and sometimes even fainter. Lunar occultations of radio sources are also of interest – remember the first quasar, 3C.273, was discovered as the result of an occultation.

Recently occultations of stars by planets (including minor planets) and satellites have aroused considerable attention.

The exact timing of such events gives valuable information about positions, sizes, orbits, atmospheres and sometimes of the presence of satellites. The discovery of the rings of Uranus in 1977 was the unexpected result of the observations made of a predicted occultation of a faint star by Uranus. The duration of an occultation by a satellite or minor planet is quite small (usually of the order of a minute or less). If observations are made from a number of stations it is possible to deduce the size of the planet.

The observations need to be made either photoelectrically or visually. The high accuracy of the method can readily be appreciated when one realizes that even a stop-watch timing accurate to $0^s.1$ is, on average, equivalent to an accuracy of about 1 kilometre in the chord measured across the minor planet.

Comets in 1992

The appearance of a bright comet is a rare event which can never be predicted in advance, because this class of object travels round the Sun in an enormous orbit with a period which may well be many thousands of years. There are therefore no records of the previous appearances of these bodies, and we are unable to follow their wanderings through space.

Comets of short period, on the other hand, return at regular intervals, and attract a good deal of attention from astronomers. Unfortunately they are all faint objects, and are recovered and followed by photographic methods using large telescopes. Most of these short-period comets travel in orbits of small inclination which reach out to the orbit of Jupiter, and it is this planet which is mainly responsible for the severe perturbations which many of these comets undergo. Unlike the planets, comets may be seen in any part of the sky, but since their distances from the Earth are similar to those of the planets their apparent movements in the sky are also somewhat similar, and some of them may be followed for long periods of time.

The following periodic comets are expected to return to perihelion in 1992, and to be brighter than magnitude +15.

Comet	Year of discovery	Period (years)	Predicted date of perihelion 1992
Chernykh	1977	14.0	Jan. 27
Giacobini-Zinner	1900	6.6	Apr. 13
Grigg-Skjellerup	1902	5.1	July 22
Shoemaker (2)	1984	7.9	Aug. 6

Minor Planets in 1992

Although many thousands of minor planets (asteroids) are known to exist, only 3,000 of these have well-determined orbits and are listed in the catalogues. Most of these orbits lie entirely between the orbits of Mars and Jupiter. All of these bodies are quite small, and even the largest, Ceres, is believed to be only about 1,000 kilometres in diameter. Thus, they are necessarily faint objects, and although a number of them are within the reach of a small telescope few of them ever reach any considerable brightness. The first four that were discovered are named Ceres, Pallas, Juno and Vesta. Actually the largest four minor planets are Ceres, Pallas, Vesta and Hygeia (excluding 2060 Chiron, which orbits mainly between the paths of Saturn and Uranus, and whose nature is uncertain). Vesta can occasionally be seen with the naked eye and this is most likely to occur when an opposition occurs near June, since Vesta would then be at perihelion. Approximate date of opposition (and magnitude) for these minor planets in 1992 are: Ceres, July 26 (7.5), Pallas, June 18 (9.4), Juno, December 29 (7.7) and Vesta, March 9 (5.9).

A vigorous campaign for observing the occultations of stars by the minor planets has produced improved values for the dimensions of some of them, as well as the suggestion that some of these planets may be accompanied by satellites. Many of these observations have been made photoelectrically. However, amateur observers have found renewed interest in the minor planets since it has been shown that their visual timings of an occultation of a star by a minor planet are accurate enough to lead to reliable determinations of diameter. As a consequence many groups of observers all over the world are now organizing themselves for expeditions should the predicted track of such an occultation cross their country.

In 1984 the British Astronomical Association formed a special Asteroid and Remote Planets Section.

Meteors in 1992

Meteors ('shooting stars') may be seen on any clear moonless night, but on certain nights of the year their number increases noticeably. This occurs when the Earth chances to intersect a concentration of meteoric dust moving in an orbit around the Sun. If the dust is well spread out in space, the resulting shower of meteors may last for several days. The word 'shower' must not be misinterpreted – only on very rare occasions have the meteors been so numerous as to resemble snowflakes falling.

If the meteor tracks are marked on a star map and traced backwards, a number of them will be found to intersect in a point (or a small area of the sky) which marks the radiant of the shower. This gives the direction from which the meteors have come.

The following table gives some of the more easily observed showers with their radiants; interference by moonlight is shown by the letter M.

Limiting dates	Shower	Maximum	R.A.	Dec.	
Jan. 1–4	Quadrantids	Jan. 4	15^h28^m	$+50°$	
April 20–22	Lyrids	April 21	18^h08^m	$+32°$	M
May 1–8	Aquarids	May 5	22^h20^m	$+00°$	
June 17–26	Ophiuchids	June 20	17^h20^m	$-20°$	M
July 15–Aug. 15	Delta Aquarids	July 29	22^h36^m	$-17°$	
July 15–Aug. 20	Piscis Australids	July 31	22^h40^m	$-30°$	
July 15–Aug. 25	Capricornids	Aug. 2	20^h36^m	$-10°$	
July 27–Aug. 17	Perseids	Aug. 12	3^h04^m	$+58°$	M
Oct. 15–25	Orionids	Oct. 22	6^h24^m	$+15°$	
Oct. 26–Nov. 16	Taurids	Nov. 3	3^h44^m	$+14°$	M
Nov. 15–19	Leonids	Nov. 17	10^h08^m	$+22°$	M
Dec. 9–14	Geminids	Dec. 13	7^h28^m	$+32°$	M
Dec. 17–24	Ursids	Dec. 23	14^h28^m	$+78°$	

M = moonlight interferes

Some Events in 1993

ECLIPSES

There will be four eclipses, two of the Sun and two of the Moon.

May 21: partial eclipse of the Sun – North America, Europe.

June 4: total eclipse of the Moon – western North America, Australasia, Asia.

November 13: partial eclipse of the Sun – South America, Australasia.

November 29: total eclipse of the Moon – Europe, Africa, America, Asia.

THE PLANETS

Mercury may be seen more easily from northern latitudes in the evenings about the time of greatest eastern elongation (February 21) and in the mornings around greatest western elongation (November 22). In the Southern Hemisphere the dates are April 5 (morning) and October 14 (evening).

Venus is visible in the evenings until March and in the mornings from April until December.

Mars is at opposition on January 7.

Jupiter is at opposition on March 30.

Saturn is at opposition on August 19.

Uranus is at opposition on July 12.

Neptune is at opposition on July 12.

Pluto is at opposition on May 14.

Desert Astronomers

DAVID ALLEN

The Chaco Paintings

Autumn was already well advanced in the canyon, yet the Sun shone with the fierceness of high summer in the desert. We had been walking for a couple of hours, the supply of water we carried was running low, and our thoughts inevitably drifted back to the bottles of cold drinks we had left in the vehicle. Our three children had talked continuously, it seemed, of what they would drink on their return. To conserve water in a dry climate by keeping one's mouth shut is a lesson only slowly learnt.

The journey thus far had been filled with interest. Initially we had kept close to the canyon's north-eastern walls, here about twenty metres high. The sandstone wore a dark coat of desert varnish, brown oxides of iron and manganese, into which were etched hundreds of patterns to reveal the paler rock beneath. Some were geometrical, some depicted humans and bighorn sheep, while others were the signatures and graffiti of more recent travellers.

Once across the canyon floor we had ascended a sketchy path to the ruin of Peñasco Blanco, a small settlement of oval outline built between 900 and 1100 AD. The buildings once rose four stories high and probably included more than 300 rooms, but the searing winds of 700 summers have taken their toll and the walls of Peñasco Blanco stand now little more than three metres tall at their highest. Layers of coloured stones set into the outer walls hinted that the masons had an artistic eye.

Only a little exploration was needed to reveal more art forms. Peñasco Blanco awaits archæological excavation, so its treasures lie not on museum shelves but in and on the ground. Shards of pottery littered the open plaza that typifies Chaco architecture; presumably they have lain there since drought and a failing economy drove the inhabitants out in the thirteenth century. Outside the walls my wife Carol located the midden, and here a wealth of potsherds awaited our attention. Previous visitors, ignoring the regulations, had collected specimens together into several huge piles where the many styles and ages could be compared. We should be glad, perhaps, that not all the shards had found their way into the pockets of those visitors: some of the black-and-white ware was particularly attractive (Figure 1). At length we had dragged ourselves away from Peñasco Blanco and descended into the canyon once more. A path now led to the welcome shade of its west wall, where we were desiccated more slowly by the dry air alone. We crossed Chaco Wash, no more than a couple of mudholes, and marvelled again that a civilization could have blossomed in so harsh a desert environment at a time when the climate was little different from today. Except on the rare occasions when rain falls, Chaco Wash, the major water course of the canyon, is dry throughout much of its length.

The end of the path was fairly evident. The grey stones were trodden to an unusually large patch of dust close beneath an overhang of rock. The walls here are of a pale sandstone, almost grey, testifying to more rapid erosion than in regions coated with desert varnish. Carving this rock would leave no prominent colour

Figure 1. Shards of pottery lie in the sand of an old midden.

differences, so the ancient Indian artists here were forced to apply paint to create their records. Painted works, pictograms, are much rarer than carved motifs, petroglyphs, in part because of the rarity of pastel rocks to paint on and in part because time more quickly erodes or fades any resultant art. Another factor may be that carving can be done spontaneously with a sufficiently sharp stone (witness the number of carved graffiti) whereas painting requires the preparation of pigments and provision of a brush.

Whatever the reason, the rarity of rock paintings around Chaco Wash encourages their examination in some detail. In this case the subject is of more than passing interest, to me at any rate. Indeed, this spot is rapidly becoming a Mecca for travelling astronomers.

Above us as we stood on the patch of trampled dust the pale canyon wall rose vertically for somewhat more than four metres, then stepped out for a metre or so in an overhang before continuing its journey upward to the sky. The paintings of interest occupied this overhang. Here, protected from the elements, they have retained a remarkable freshness over the centuries since they were executed.

Three symbols there are, each about the size of a dinner plate, each painted in brown. On the left, closest to the canyon wall, is a ten-pointed star. To its right a crescent moon has a slightly darker hue and may therefore have been painted at a different time. Above, near the outer edge of the overhang, a left hand has been painted. The hand is a common feature of rock art here, and is thought to represent the artist's signature; the star and the moon are almost certainly astronomical (Figure 2).

The now popular interpretation can be no more than conjecture. Had you stood at this spot, with your back to the rock, at dawn on the 4th day of July 1054 AD, you would have been treated to the extraordinarily rare sight of a bright supernova rising just a little to the right of a crescent moon. One day earlier the supernova would not have been visible, for it burst on the scene with that startling suddenness that among celestial phenomena only comets, novæ and supernovæ achieve. The 1054 supernova became as bright as the planet Venus, outshining all other stars, and remained visible for several months. Did some unknown Indian astronomer, at the peak of the Anasazi culture, observe that striking event and record it on this very overhang while the memory was fresh in his mind? We cannot know until archæology finds a way to date the paintings with sufficient precision, but it is an attractive and even a credible tale.

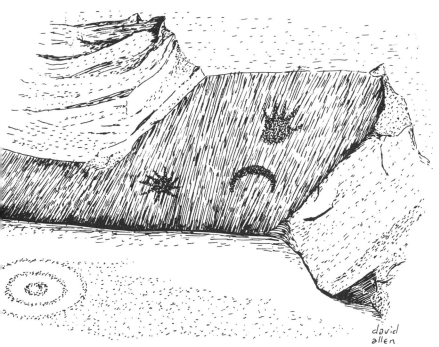

Figure 2. Looking up under the overhang at the astronomical rock-art site.

On the rock face below the overhang another symbol has been painted. Using red and yellow pigments the artist produced a series of concentric circles of alternate colour. Astronomers are keen to attribute all the paintings hereabouts to celestial events, and archæologists make it an easy task by pointing out that concentric circles are thought usually to symbolize the sun. The story would end there but for the sharp eyes of a park ranger who had suggested to me that there was more to this design than meets the casual eye. True enough, a faint redness stretches out to the right of the concentric circles, tapering slowly to a broad terminus something like 1.5 metres away. Is this natural pigmentation in the rock or a badly faded coat of red paint? To my knowledge no study has been made of the red patch, and having no ladder with us we were unable to make a close inspection to form our own views. The similarity of hue to that of the red circles does suggest paint, and if that is the case the whole pattern resembles not so much the Sun as a comet.

If the Chaco astronomers did record a comet here at their

astronomical diary rock, it would presumably have been a pretty bright specimen, and one that came by near the peak of their culture, around 1000–1200 AD. One of the brightest of that period in fact appeared only twelve years after the supernova. We know of it as Comet Halley, making that famous 1066 visit that worried the Saxons and encouraged the Normans in their conquest of Britain. There's no way to prove that this is a representation of Halley's Comet, but I find it an intriguing idea.

The Anasazi

The Chaco Culture National Historical Park is arguably the most important archæological site in the USA to pre-date European influence, and certainly one of the most spectacular to visit. The ruins had impressed us greatly, as had the information provided at the park's visitor centre in the form of leaflets, books, and films. It was unquestionably worth the long drive over the potholed and corrugated sand roads of northern New Mexico to reach the place. Here, at their cultural peak, about 6000 desert Indians dwelt. Here they constructed their massive cities of terraced apartment blocks; here they farmed corn, beans and squash; here they worshipped their pantheon of gods and goddesses in giant subterranean temples called *kivas*; and from here, shortly before the year 1300, they vanished without trace (see Figure 3).

Nobody knows what these people called themselves. When Europeans arrived the region was inhabited by a small population of Navajo farmers. The Navajos had migrated from northern Canada long after the ruins were abandoned. It is by a Navajo word that we know the people of Chaco Canyon – *Anasazi* – 'the departed ones'.

Extensive archæological work has revealed the true extent of the Anasazi culture, covering an area larger than New Mexico itself. Two major groups existed, of which the Chaco tribe was the more influential. Only half of the Chaco population lived at Chaco Canyon. A comparable number occupied smaller towns scattered hundreds of kilometres away across the desert. Broad roads linked all Chacoan towns to the city complexes of the canyon, roads wide enough for juggernauts to pass one another, though they carried only foot traffic. Chaco Canyon was the administrative centre of the region, the head office. Food and other resources were shared through Chaco so that if, say, drought affected the southern towns then their larders could be swelled by the excess from another region. Like any administrative centre, Chaco Canyon grew to be

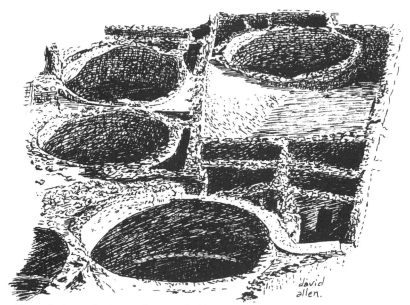

Figure 3. Anasazi ruins are distinguished by the circular, subterranean rooms known as kivas. *Each served as the religious centre for an extended family, but was used almost exclusively by males.*

the biggest part of the whole, and as a result the least adaptable. When a prolonged drought affected the region in the late thirteenth century, the cities of Chaco were abandoned. Gradually the other sites closed too, and the population somehow melted away into the desert.

That, at least, is the story that is often told. In reality the population would have moved on to settle other regions or, more likely, to join with their relatives living where the rains were more reliable. The majority probably moved down to the Rio Grande river in Texas, where their culture changed somewhat. In some parts of the Anasazi range, however, the towns were not abandoned. We were later to include in our tour the Hopi Indian town of Old Oraibi and the better-known Acoma in northern New Mexico. Both were built about 1150 AD and vie for the distinction of the longest continuously inhabited site in the USA. Our experience of Chaco and many of the other Anasazi sites gave emphasis to our visits. Indeed, to enter either of these towns in many ways was to travel back in a time machine to Chaco's prime.

From today's Pueblo Indians, as both modern and ancient town-dwelling tribes are collectively known, archæologists gather their clues to interpreting the ruins. When it comes to Anasazi astronomy, therefore, we have at least a first hint of what to expect. That hint certainly allows the possibility that the Anasazi kept a sufficient watch on the sky to notice both comets and supernovæ. It also encourages us to expect that a reasonable calendar was kept, not just to regulate the planting of crops, but also for the observance of regular religious ceremonies. Some of the buildings at Chaco Canyon and elsewhere do seem to include alignment windows that might have indicated at least approximately the summer or winter solstice, though archæologists are divided on whether they were ever used as such. But the most remarkable astronomical remains were not the sort of thing anyone would think to go out and look for, but features that took everyone by surprise.

Pecked Spirals

In the north of Britain there are many examples of prehistoric rock carvings whose significance is utterly obscure. The cup-and-ring markings feature a hollow about the size of half a tennis ball surrounded by a shallower concentric annulus. No sensible explanation has, to my knowledge, been proposed for the cup-and-ring markings. Almost identical carvings exist in profusion on an expanse of old lava on the east coast of the island of Hawaii. The Hawaiian carvings might be equally puzzling but for an oral legend that describes their use. New mothers reputedly could earn good fortune for their children if they burnt the umbilical cord in one of these hollows.

Now, I'm not suggesting that the British cup-and-ring markings had similar significance. Rather, my point is that the most brilliant archæologist could neither happen upon such an interpretation nor persuade others of its veracity with only the meagre clues that time leaves. Another prime example of such unpredictability has been bequeathed to us by the astronomers of Chaco Canyon.

Panels of petroglyphs, carved or pecked rock-art, frequently incorporate tightly wound spirals of five to ten turns. What might have been the significance of a spiral to the Anasazi? Some have suggested snakes, though the lack of a head makes this somewhat unlikely. Others favour more esoteric interpretations. In 1977 Anna Sofaer, a young archæologist, was making a study of the spirals in and around Chaco Canyon. In the heat of a summer morning she had made the tricky ascent of Fajada Butte, an isolated outcrop of the canyon walls at the southern end of Chaco Canyon (see Figure 4). Just below its summit a pair of well-engraved spirals was known, in a sheltered spot beneath a slab of rock that had slid down from the crags above and had split into three portions. Morning passed into afternoon as she sketched and noted the details of the spirals, when suddenly she became aware of a narrow vertical line of sunlight coming through one of the cracks between the rock slabs. The line of sunlight descended quickly as the Sun moved round the sky, passing near the centre of the spiral before it vanished again into shadow at the bottom.

Figure 4. The seemingly impregnable Fajada Butte forms a striking silhouette after sunset.

To cut short a major epic, what Anna Sofaer had stumbled upon was a calendar marker. Subsequent analysis has shown that the triple slab of rock, each piece of which weighs several tons, had not slid from above but had been moved from some distance away by the Anasazi and stood together deliberately so as to project stripes of sunlight through the gaps between. The spiral had then been carved in such a place and of such a size that on the summer solstice the dart of sunlight passes exactly through its centre (Figure 5), while at the winter solstice two darts exactly touch its outer edges. A smaller spiral, to the left, is split by a dart of sunlight at the equinoxes. At night the spirals could also have been used to track the Moon's movements.

Again, nobody could have guessed the significance of the Fajada Butte spirals, which would have remained unknown to us had Anna Sofaer chosen a cooler time of day to make her study. My only

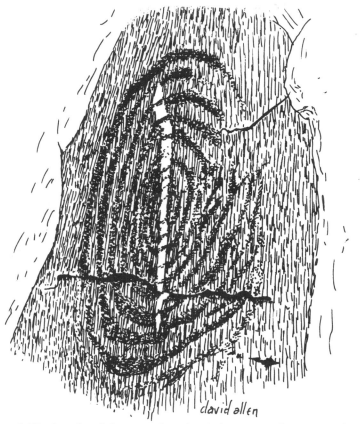

Figure 5. The dart of sunlight crosses the main spiral at noon on the summer solstice. The view is from outside the rock slabs set up to create the shadow.

regret in visiting Chaco Canyon was that Fajada Butte is out of bounds to all visitors – even to astronomers.

Holly

If Fajada was out of bounds, then we should turn our wheels to Hovenweep. A long day's drive away on poor roads, Hovenweep is an outlying complex of ruins of the Chaco culture. I couldn't have made such a journey on foot carrying my own provisions, and my admiration for the Anasazi grew once more on the drive. The dwellings at Hovenweep are small and primitive. Rather than apartment complexes, the Indians here built cramped towers, many in quite inaccessible places, suggesting that warfare was a factor in their lives. No more than 150 could have inhabited the Holly group, one of the scattered locations in Hovenweep National Monument. Most stand on the rim of a shallow canyon, but some occupy huge boulders that in prehistoric times had broken off the rim and tumbled into the canyon.

Near the Holly site is a particular profusion of boulders, each three or four metres on a side. With care one can pick a route down between them to an overhanging boulder face bearing some rather faded petroglyphs that include two spirals. Again these spirals mark the passing seasons. In this case the dart of sunlight is horizontal and is formed at dawn, seemingly by the chance alignment of various of the boulders. The midwinter Sun doesn't reach the rock face at all, but the summer solstice and the equinoxes can be read from the position of the dart of light with respect to the spirals.

I formed the impression that the Holly site was the older one, a chance shadow pattern taken advantage of by the Anasazi time-keepers. Fajada, it seemed to me, was a later and more sophisti-cated construction probably inspired by the Holly calendar. These are the only two known to date. There are many spiral petroglyphs, and the vast majority certainly are not involved in timekeeping. Many are yet to be adequately studied, however. We are only beginning to scratch the surface of Anasazi astronomy.

Return

The work that went into the Fajada site, shifting massive boulders with extreme precision, was far from slight. The Anasazi, just like so many modern nations, expended a small but not insignificant fraction of their resources into astronomy. Whatever their motives, they must have been sky watchers. Thus wandered my thoughts as we toiled back from the supernova site to the vehicle, while a late afternoon sun burnt into our backs. As we sipped the last drops from our water bottles I admired once again the Indians who had carved a civilization out of this valley. Architects, artists, scientists, defeated only by a climate that could pile up more against them than their resources could defy. Yes, I reckoned they were capable of watching and knowing the sky and spotting the 1054 supernova, a phenomenon that has left no record in the annals of the whole of Europe.

Tired but satisfied we opened the vehicle and downed several icy glassfuls.

Guide to
Stellar Photography

JOHN FLETCHER

My interest in deep-sky astronomy, and the need to view fainter than fifteenth magnitude stellar objects, requires a telescope with a very large aperture. For example, a typical large-aperture amateur telescope would be, say, a 50-cm reflector. The theoretical threshold of such an instrument used visually would be around the 16.5 magnitude mark. It is interesting to note that the Mount Wilson 2.5-m reflector has a theoretical threshold of around magnitude 18.8.

However, astrophotography carried out using amateur-sized reflectors can reach limiting magnitudes that are well below the limits of visual observation aperture for aperture. In a 20-minute exposure using my 25-cm (10-inch) reflector at Newtonian focus, using fast Tri-x film, I have recorded stellar images to 18 mpg (photographic magnitude). It is important to note that one must not put a guess on limiting magnitudes. Professional limiting magnitude charts are available. The charts show standard star magnitudes (stars that are not variable). The chart (see Figure 1) I used for comparing my negatives in one instance was an area of sky centred around the most remote globular cluster in our Milky Way Galaxy, namely, NGC 2419 in Lynx.

With these thoughts in mind I decided that recording deep-sky objects with the possibility of a scientific contribution, or the ultimate, a discovery, was to be my goal in life.

After reading various brochures about commercially produced telescopes and being horrified by the sizes of the larger telescopes, and the high prices required to purchase such instruments I decided to go for a 21.6-cm (8.5-inch) f6 Newtonian reflector mounted on a German driven equatorial. This was to become my first successful photographic instrument.

What is desirable, however, is a permanently housed telescope if the instrument is to be used for photographic application. This

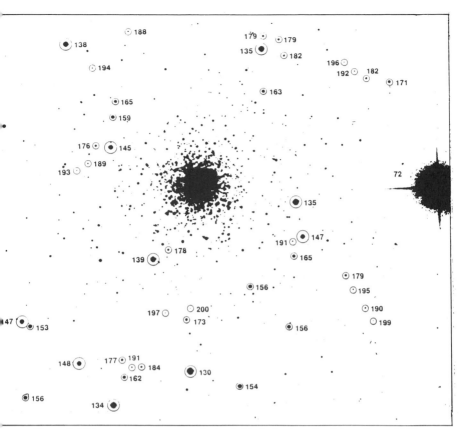

Figure 1. This chart shows star magnitudes (with decimals omitted) in and near NGC 2419 in Lynx. Photograph this cluster, then compare the faintest stars on your negative with those here. (Reproduced courtesy of US Naval Observatory.)

is where painstaking dedication and hard work came into things.

My first two observing huts consisted of low sheds with run-off roofs. These improved things immensely. Later in 1981 I built a wooden, octagonal, rotating observatory which was later to be named Mount Tuffley Observatory by members of the Cotswold Astronomical Society who learned of me through my astronomy articles in the local newspapers.

The 21.6-cm (8.5-inch) reflector arrived and I was all set for polar aligning the instrument. Luckily, unlike many commercial instruments, it came equipped with polar axis fine adjusters for both altitude and azimuth. Polar alignment is an important feature required for successful and easier guiding on the stellar image chosen. Without some means of these fine adjustments one can run into serious trouble resulting in poor image quality and/or trailed stars on one's negatives.

My most successful achievement using the 21.6-cm (8.5-inch) reflector at the Newtonian focus was a thirty-minute guided exposure using Tri-x 400 ASA film to record the approaching 14 mpg Halley's Comet in Orion. The comet was at a distance of 200,000,000 miles. I was honoured to be the fourth UK amateur to record the comet's image on film. The unseen (visually) position was located using driven setting circles. Then the accuracy of the setting circles was checked by offsetting in minutes of arc from a bright star to either side of the comet's position in the sky. The photograph (Figure 2) was secured at 03.30 UT, just two hours before setting off for work as a postman in Gloucester.

However, as I wanted a larger telescope I had to think of a larger observatory. I am very lucky to have an engineer for a cousin. After lots of careful thought we decided to go for a proper dome. With skilful planning he worked out how a dome shutter could open 14 inches past the zenith without coming into contact with the sides of the lower structure. This was done by having a zenith trap which is in the lowest frontal position of the shutter when the dome is closed. It can be independently removed before sliding the main shutter open, and later proved to be useful when viewing or photographing low altitude objects, and provides a little extra protection from the elements. I will give you a full run through of the work and time involved in building the new 10-ft 4-inch all-metal dome.

The New Mount Tuffley Observatory

We started work on the design on January 5, 1986, and by January 25 had completed the skeleton. My first task was to lay a 12-foot square concrete base, which I mixed by hand. Several coats of red oxide paint were put on the dome skeleton to protect the metal. In order to work on the dome I rested it on loaned milk crates in my garden. After ordering 14 sheets of flat 20-gauge aluminium alloy, 2000 rivets, a hand rivet-gun and four small G clamps, I started to do the most soul-destroying and tedious job I have ever done in my entire life. On February 3 I started drilling out some 1700 holes through the quarter-inch thick, metal-rib structure. For many hours every day for ten days, in appalling winter weather, I proceeded to drill the holes some 1½ inches apart. Then came the cutting and shaping of the alloy, followed by hand riveting some 1700 rivets, and using a flexible sealer to get weatherproof joints. It was March

Figure 2. Photograph of the distant 14th magnitude Halley's Comet.

28 of that year when the dome was completed. The first two weeks of April I spent putting etching primer and five coats of matt grey enamel coach paint on the dome. It then stood complete and was a very rewarding sight to see. After this I built a square structure for

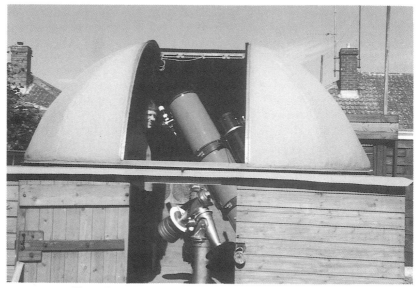

Figure 3. The new Mount Tuffley Observatory, August 1987.

the dome rail to sit on which was made of timber and stood 4 feet high. The advantages in having a square, lower section is that when inside the dome one has lots of room for storage under the four corners beyond the diameter of the actual dome. I then sealed the concrete floor against dampness, fitted the thrust and roller bearings to the circular track, and with the help of eight adults the dome was lifted into place. The next job was to put in a raised wooden floor which has a fitted carpet to keep the feet warm. This floor is totally separate from the centre position where the present 25-cm 6.3 reflector stands. Finally after some 300 hours labour the insides were lined with marine ply to make things look very professional and also help prevent any dampness when the equipment is not in use (Figure 3).

Getting started with Astrophotography

Most commercially produced reflecting telescopes, although still expensive are quite suitable for visual work. At the same time many fall very low on my list if they are to be used for photographic application at, as in my case, the Newtonian focus (see Figure 4).

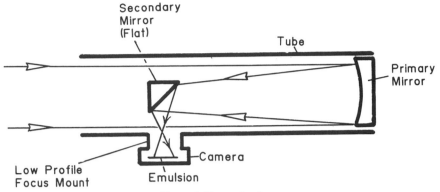

Figure 4. Newtonian focus.

The accuracy needed to set up a telescope becomes more critical as the focal length of the instrument becomes greater. Many factors come into this as I was to find out when setting up my new arrival of a 25-cm (10-inch) f6.3 reflector, a commercially produced instrument. With my own determination and technical guidance from two close colleagues (and notable fellow amateur astrophotographers, namely, Bernard Abrams and Ron Arbour) I was able, after nearly nine months of making alterations and adjustments to the equipment, to carry out the 3.5 seconds of arc guiding tolerance required to record sharp stellar images. This allowed me over a period of time to gather a huge collection of acceptable pictorial photographs (see Figures 5–8) and obtain the faintest limiting magnitudes possible (when required) for recording faint quasars, BL Lacertæ objects, searching for and recording the rise and fall of supernovæ and novæ outbursts.

Things to consider

When ordering a reflecting telescope to be used for Newtonian focus astrophotography there are many things to consider. One thing needed was a larger secondary mirror to collect and direct the entire light cone received from the primary mirror. A standard 50-mm minor axis, flat mirror or smaller is supplied with most 25-cm instruments. My reflector was fitted with a 76-mm minor axis secondary on purchase, the size needed to obtain near maximum light-gathering power for this size of instrument when used photographically. Even so, unfortunately the camera body of the 35-mm camera will obstruct some of the incoming light. For those wanting

Figure 5. Messier 13 Herculis taken with 25-cm (10-inch) f6.3 reflector, 5-minute exposure on Tri-x. (John Fletcher)

to make improvements on this they must make modifications to their 35-mm camera or purchase a different type of camera, preferably from a specialist.

Another thing is to ask for a larger, say 50-mm (2-inch) diameter,

Figure 6. Double cluster in Perseus taken with 25-cm (10-inch) f6.3 reflector, 4-minute exposure on T Max 400. (John Fletcher)

low-profile rack-and-pinion mount be fitted. The reason you need a larger diameter rack mount than the standard 38-mm (1¼-inch) one is that the light entering the camera from the secondary mirror is cut off and severe vignetting at the corners of the

Figure 7. Quasar 3C273 VIR taken with 25-mm (10-inch) f6.3 reflector, 3-minute exsposure on T Max 3200. (John Fletcher)

35-mm negative is experienced.

Also it may be necessary to move the main mirror with its cell, up the tube towards the secondary mirror to obtain focus through the

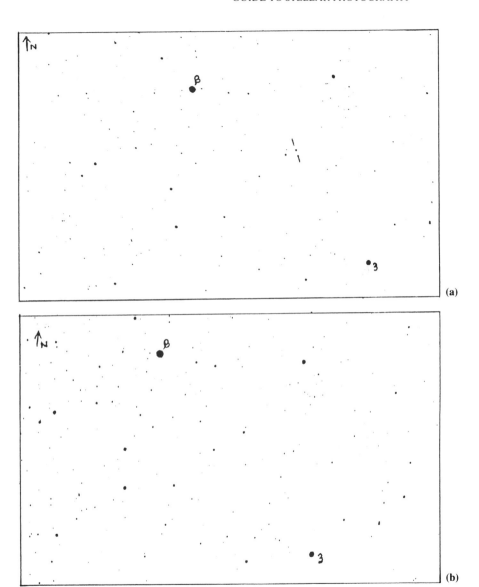

Figure 8. Two photographs of Nova-Vulpeculæ I taken during an unguided search for novæ using a 300-mm, f5.6 lens, piggy-back mounted. (a) Taken 21.30 UT, 20/8/84. (b) Taken 21.20 UT, 28/8/85. (John Fletcher)

camera. This is a common problem not spoken of until you discover it for yourself. After this alteration I had an extension tube made so that I was able to focus eyepieces for visual observations.

Most deep-sky photography is carried out using short-focal ratio reflectors of say f4.5 or f5.0. Using such short-focal ratio mirrors as these, collimation (alignment of the optical path) is a critical factor. A tip that was given to me many years ago was to mark a black spot (a few millimetres in diameter) on the centre of the main mirror. First make sure that you air-blow any possible dust specks off the surface of the mirror to avoid tiny scratches. Then, to determine the centre of the main mirror, place a disk of the correct diameter, made from soft white blotting paper, cautiously on to the surface of the mirror. The disk should have a small pinhole in its centre. Using a soft black felt-tipped pen of the water-based type, mark the centre through the hole. From then on, when using the collimating eyepiece, you will have a reference point for aligning all the optics very accurately. As the black spot is at the centre of the mirror there is no extra light loss.

Another thing to look out for when the camera is in focus is the inner end of the rack-and-pinion draw tube. If its bright chrome barrel is protruding into the tube, cut it off! It cuts into the narrow light cone entering the rack mount, and often causes defects in the images of stars recorded on film.

Polar Alignment

The first problem encountered was that the 25-cm telescope had no means of fine azimuth and altitude adjustments. The only way I was able to make any adjustments was to move the entire telescope. Including the 11.4-cm (4½-inch) reflector guidescope, counterweights, and camera attachments, the telescope weighed well over 200lb. With the aid of friends the 200lb-plus sitting on the equatorial head was held steady whilst the main central locking was undone. By moving all this weight up and down in altitude on the loosened main head bolt we were able to get the altitude right to within half a degree of the true pole. The main equatorial head bolt was then locked. To try to turn such weight manually on an equatorial head was an impossibility if accurate alignment was to be achieved. Many experienced amateurs can, however, obtain arc minute accuracy eventually, but a higher degree of accuracy is needed for all but the very short focal length instruments. How I overcame this problem was very unconventional indeed but cheaper than any professional

GUIDE TO STELLAR PHOTOGRAPHY

alterations which would also have meant being without the equipment for several weeks or much longer, hence no astrophotography.

First, prior to collection, the supplier was asked to drill and tap a vertical hole into the end of one of the three telescope-pedestal mounting legs. A threaded three-quarter-inch bolt was fitted. This leg was to point north and the bolt was to be the fine adjustment for altitude and also to act as a pivot point for azimuth whilst fine adjustment was being made at one of the other pedestal legs. From then on I was able to adjust the vertical bolt accordingly to make fine altitude adjustment.

Now what about Azimuth?

First, I put a heavy duty metal plate under the other two pedestal legs and fitted a horizontal push-and-pull bolt on either side of one of the legs for effecting fine azimuth adjustments. Once again, approximate azimuth adjustment had to be made manually first to within half a degree of the pole. My idea was that after manual set-up, by turning the adjustments the whole mount and telescope together could be moved east or west to enable arc second accuracy to be gained in azimuth.

To correct an azimuth error the star drift method is best. Orientate the cross hairs of a guiding eyepiece NS, EW and select a star close to, and east of, the southern meridian, and at 0° declination. The star observed when east and again when west of the meridian will show any azimuth error. A northern drift tells you that the axis is west of the pole, a southern drift, east of the pole.

Things Move in the Night

The mechanics of the actual mount and telescope assembly almost always flex under their own weight as longer exposures are carried out.

PVC and other man-made materials used for telescope main-tube assemblies flex with photographic guidescope and camera weight and temperature as well as from the relative positions of the instrument when in use (even when the instrument's balance is correct). Main mirrors can move, as can other parts, causing everything but points of light on film. This is often so, even when the guide star has been kept dead-centre within the guiding reticle throughout an exposure.

In the manual that accompanied a brand-new telescope the manufacturer stated that one must not worry if the main mirror is

143

heard to move in its cell as a knock when in use because it should do so. This telescope was purchased with a Newtonian focus adaptor. If a long exposure were to be made displacement of the image would be likely to occur with so much movement between mirror and its cell. Throughout the entire system a shift of one-thousandth of an inch can often cause a displacement of the stellar image.

One way of checking for any flexure is to use two high-power guiding reticles at the same time. Place one reticle in the main instrument and the other in the guiding telescope. Choose a third or fourth magnitude star and place the same star at the centre of each reticle's cross hairs. Now after twenty to thirty minutes of guiding with one telescope, and keeping the star central, look into the other telescope, and if that same star is not in the centre of its cross hairs there has been flexure or a shift of either the optics or the mechanical structure of the telescope. Guidescopes are often the main cause of this. They are often held in place with adjusting bolts set in two circular metal rings with plenty of space between them and the actual guidescope, allowing large areas of sky to be selected around the object. It is important that a guide star should be as close as possible to the object being photographed.

The message is that polar alignment and any periodic errors have to be ironed out to make way for better and easier guiding and good image quality. Constant adjustments and minor alterations are needed until you iron out the worst of the problems.

In Britain, with its temperate climate, I have been all set for a long night of deep-sky astrophotography only to discover that the atmospheric seeing was so poor that guiding was impossible. Using my illuminated guiding reticle I guide on a power of ×450 so any errors or possible run offs needing correction are noticed and corrected more often than is possible using a lower- and under-powered guiding system.

I have made approximate measurements of poor seeing on several relatively clear nights and found stars to be moving around the guiding reticle cross hairs as much as 8 seconds of arc.

Astronomy with a 25-cm Telescope

One type of astrophotography I carry out is by not actually using the main instrument but a 300-mm f5.6 to f32 Optimax telephoto lens which I purchased for £10 in 1984 and have used ever since. I have found it ideal for nova patrolling and recording large deep-sky objects on film. It records a field of approximately $7° \times 4.5°$ and its fastest setting of f5.6 gives surprisingly good stellar-image quality even to the edge of the field. Stationary exposures of only 3 seconds are possible before the diurnal motion of the Earth will cause the star images to trail. However, a 300-mm lens when mounted on a well-aligned, driven telescope (piggy-back mounted) things are quite different. Without any drive correction one should be able to open the shutter for, say, 20 minutes and record perfect points of light.

My main 25-cm f6.3 instrument is used for both pictorial photography and capturing passing comets on film.

I am also interested in the BAA active galaxy programme, and produce finder charts to help others locate more easily the faint quasars and BL Lacertæ objects. However, to carry out serious photographic work and study their possible magnitude fluctuations one must use spectroscopic series film such as 103A-O (blue sensitive) in conjunction with a professional 'B' filter whose transmission peaks at 433.0 nm (nanometre). The stellar diameters can then be measured against standard stars using a machine called a densitometer. Prior to this, however, the unexposed film and sky-fog levels must be measured on the plate or negative with a machine called a sensitometer. Without sensitometry and densitometry accurate measurements are impossible.

Much more accuracy on this type of work is achieved by amateurs doing photometry, but not at the low light levels of many quasars. (See Figures 9 and 10.)

My main interest though is the search for supernovæ outbursts in external galaxies. I use the main instrument at the Newtonian focus and aim for a limiting magnitude of around 16.5 mpg. For this I use Tri-x Professional 400 ASA film, or on transparent nights T Max Professional 3200 ASA film. These films are developed on the same morning as taken, prior to retiring for the day or setting off for work. The films are developed using D19 at 20°C for 5 minutes with the Trix and 4 minutes for the T Max film. Each galaxy has to be photographed twice so as to eliminate minute dust specks or flaws on the emulsion which may cause false alarms.

Checking the negatives is the last job and I use a home made blink comparator. Using a previous negative (my own master negative) and the new one taken of the same field I place them on two illuminated diffusers. Then looking down through two separate magnifying lenses with both eyes I superimpose the views of them. I

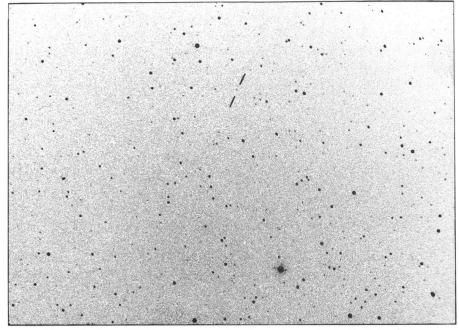

Figure 9. Quasar 3C66a in Andromeda using a 25-mm (10-inch) f6.3 reflector, 8-minute exposure on 3M1000. (John Fletcher)

am able to turn one of the diffuser's light on and off. In doing so a new star would appear to blink on and off.

An alternative way of checking new negatives against old is to set up two slide projectors side by side and do a star comparison around the galaxy in question.

Slide Duplication

Finally, a device that I find most useful is a slide duplicator. When you duplicate a colour slide on to a finer grain ordinary colour film there is an increase in contrast. The film I use to copy on to are, for colour Kodachrome 64 ASA and Kodak Ektar 25 ASA print film. In all copying one must bracket exposures a few stops either side of the meter reading of your SLR camera to obtain the most pleasing results. I have used a slide projector light to illuminate the duplicator and have found that the colour balance shifts towards the red. Some nebulæ such as the California Nebulæ or the Rosette Nebulæ may improve immensely this way. Other subjects, however, look

Figure 10. BL Lacertæ object, Markarian 421, taken with 25-mm (10-inch) f6.3 reflector, 4-minute exposure on Tri-x 400. (John Fletcher)

unnatural with more red in the picture. One can overcome this shift to the red by using sunlight or an electronic flash.

Also by copying black and white negatives on to a fine grain film such as Tec Pan 2415 ASA one can produce positives. From these positives you can then produce negative prints with greatly added detail, especially in the case of the faint spiral arms of galaxies. In addition, the duplicator can be used as an enlarger when copying.

My astrophotography has had lots of pitfalls compared to visual work. These, however, were offset by being able to record very faint stellar objects. Also, due to the human eye's poor response to colour, the faint starlight and nebulæ seen are often not very impressive – but time exposures of the same objects recorded on film are magnificent. It is this that motivates me and many amateur astronomers to photograph the night sky and I feel we will continue to do so for many years to come even though much more can be learned these days by what I call exciting, but unromantic, 'Electronic Astronomy'.

The Planets of Interstellar Space

PAUL MURDIN

Our Earth is a planet, and, like the other eight of our Solar System, it is bound by the force of gravity to our star, the Sun. The Sun is one star of a collection called the Galaxy.

In the *1991 Yearbook of Astronomy* I described how stars, although usually collected into galaxies, can sometimes be found between the galaxies, in intergalactic space. If stars exist which are separate from galaxies, can planets exist which are separate from stars? Are all planets attached to a star, or are there planets which wander isolated in interstellar space?

We know that from Pluto, the most distant planet of our Solar System, the Sun appears greatly diminished in brightness, more akin to the cold brightness of full moon than of the warming, life-giving, comforting blaze which we on Earth experience during the day. But imagine a world without a sun at all, coursing in perpetual darkness, with only starlight to illuminate its frozen wastes – is this a scene entirely from our imagination, or are there actually worlds like it?

A plain answer is that astronomers do not know if such worlds exist or not. There is no technology by which to detect them.

Astronomers suspect – but are unable to affirm for certain – that there do exist planets which circle other nearby stars. Such a planet orbiting a nearby star could betray its presence by its pull on the star. It would shift the star first to one side and then to the other side of a straight-line flight through space. There are some stars in which this behaviour has been suspected by astrometrists from Earth-based observations and the more accurate measurements from the Hipparcos satellite (which in 1991 is likely to be nearing the end of its mission in space) may be able to confirm that planets, other than our own, exist – around *some* stars.

Astrometrists' colleagues, the infrared astronomers, have suspected that there are planetary systems orbiting some nearby stars, because they can detect the presence of planetary dust, warmed by the light of the parent star and emitting infrared radiation. The InfraRed Astronomy Satellite (IRAS) has observed such disks around the stars Vega and Beta Pictoris.

One hope for the Hubble Space Telescope, after its optical deficiencies have been corrected by additional lenses installed in a future shuttle flight, is that it may be able with its increased resolution to view the glint of a planet, reflecting the light of its parent star.

No data of any of these kinds could be determined from an isolated planet – there is no star nearby to an isolated planet which the planet can affect by gravitational force, nor any star whose light it could reflect or re-radiate as infrared. So the bald answer to the question: do isolated planets exist? is 'don't know'.

However, Martyn Fogg, a London-based amateur astronomer, has done a bit better than this. In a little-known article in *Earth, Moon, and Planets* (volume 43, page 123, 1988), Fogg attempted to calculate the number of isolated planets in our Galaxy. Naturally there is considerable uncertainty, but Fogg estimated that in the solar neighbourhood there could be one isolated planet for every 30 to 3000 stars, and in the galactic bulge the figure might be as much as one isolated planet for every ten stars.

Where might these planets come from? It is in attempting to answer this question that Fogg has been able to frame his calculations. First Fogg distinguishes between small stars and planets, following NASA astronomer David Black.

According to Black, there is a sharp line between brown dwarfs and planets. Brown dwarfs are stars; they form by fragmentation of a gas cloud but are not large enough to heat their inmost parts to temperatures at which nuclear burning can be sustained. The only source of energy which brown dwarfs have (after a brief period of deuterium burning) is gravitational settlement. This warms them, and they emit rather feeble amounts of infrared radiation. Several claims that brown dwarfs have been detected have been made as the rapidly developing infrared technology has been applied by astronomers to surveys of stellar nurseries where bigger stars have formed, have switched on nuclear burning, and are making themselves obvious by the light which they emit. The minimum size for a nuclear-burning star is around 0.08 solar masses (one twelfth the mass of our Sun). It seems that brown dwarfs themselves have a minimum mass of about one hundredth the mass of the Sun, this being the size of the smallest lumps in the fragmentation of an interstellar cloud.

By contrast to brown dwarfs, which form by fragmentation of a gas cloud, and are all of the same composition, planets form in the disk of material produced around a newly formed star. In the process of contraction, the newly born star spins faster. The newly born star dissipates its rotational speed and produces a disk. The disk rolls together into lumps which accrete matter from nearby in the disk, and planets grow – they have chemical compositions and masses which depend on distance from the centre of the disk. To judge by the smallest and the largest planets of our Solar System (Mercury and Jupiter), planets can lie in the mass range one ten-millionth to one thousandth of the mass of the Sun, much smaller than brown dwarfs.

According to this view, planets never form as isolated masses – they must form near a star. If there are isolated planets this must be because the connections between the planets and their parent stars have become severed. To emphasize the origin of the interstellar planets, Fogg calls them 'unbound planets'.

One way in which planets may be shed from a star is by the intrusion of a second star into the planetary system. Obviously a star which moves through a system of planets will greatly disturb them but what exactly happens depends on the circumstances of the encounter. A star which plunges perpendicularly through the plane of a planetary system has less of an effect than one which follows a trajectory within the plane. A star which meets a planet nearly head on has a shorter time in which to affect its orbit than a star which overtakes a planet.

J. G. Hills of Los Alamos National Laboratory studied by computer 40,000 encounters of a star with a planetary system like our own. He found that when a solar-type intruder into a planetary system passed within a distance of two or three times the distance of a planet from the Sun, the encounter tended to move the planet outwards in the Solar System and even disrupt the planet entirely from the Sun. Close encounters which did not lead to disruption of the planetary system usually resulted in the intruder carrying off the planet.

It is interesting to speculate on what the inhabitants of the planet would experience. Throughout the history of civilization on the planet, the solar-type intruder star would be a naked-eye star, which would gradually grow in brightness as it tracked towards the planet and its sun. In the last millennia before the encounter, the intruder would increase in brightness to outshine the naked-eye stars, and in the last century would rival the full moon. The gravitational disturbance would be calculable at first from observations of the outer planets of the planetary system, but in the last few revolutions of the planets (last few years), there would be major departures of the orbit of the planet from its regular course, enough to cause changes of climate. Over a time of a few months, the planet would be broken free from the gravitational pull of its sun and course outwards from its solar system, probably curving high above its former orbital plane and pulling ever further from its sun over several years. The initial disturbance is likely to heat the surface of the planet to unbearable extremes, but afterwards everything would freeze, as the planet gradually receded from its sun and the intruder.

In our own Solar System, if an intruder passes within, say, forty astronomical units from the Sun (i.e. within the orbit of Pluto) we can expect it to have a marked effect on the orbits of the planets, at least to leave them very non-circular. Fortunately for us, this appears never to have happened – if it had, so that the Earth moved back and forth by large distances from the Sun in a highly eccentric orbit, then the annual temperature changes would have made the development of life on Earth very much more difficult, and I doubt that I would be here to write this article nor you to read it. J. G. Hills calculated that the odds against an encounter of this closeness in the life of the Sun were about six to one. The odds would shorten if the Sun were to pass through the galactic bulge of our Galaxy since stars there are packed closer together and the chances of a close encounter increase.

Another way in which the connection between planets and their star can be severed is if the star becomes a supernova. Virtually all the star's body is ejected at high speeds, and blasts past any planets which remain in orbit around it, having survived the previous evolution of the star. The planets will be heated and their surfaces scoured by the blast wave; if they survive this, then they will fly off into space, the gravitational link between them and the star having disappeared as the supernova passed outwards across their orbit.

Only stars of high mass end their lives as supernovæ – stars which are less than five solar masses evolve more quietly. It is not clear whether such high-mass stars have planetary systems. Our own slowly rotating Sun, at one solar mass, is the only star which we can be sure has planets, and the indications from studying star rotation speeds are that only stars with a mass of two solar masses or less rotate slowly and therefore seem likely to have planetary systems. This, however, may be misleading; perhaps all stars have high spin rates at first and slow down with age, not just because they have created planets. Massive stars may not have had enough time to lose their spin, so we are fooled by connecting rotation speed and the existence of planetary systems. Perhaps all stars, slowly rotating or not, have them. In that case, the massive stars which explode as supernovæ must shed any planets which they possess at the time.

Fogg shows in his 1988 paper that supernovæ are likely to be the dominant method of releasing planets into interstellar space, if massive stars have planets. In the solar neighbourhood, there could in this case be one interstellar planet for every thirty stars. If planets are released into space only by stellar encounters then there is only one for every 3,000 stars.

Even though interstellar planets may be relatively rare by this comparison, there would be, on the conservative estimate, literally millions of them in our Galaxy. They would be almost undetectable – 'dark matter'. However, they cannot account for a significant proportion of the 'missing mass' which astronomers detect by its gravitational pull but which emits no radiation. Interstellar planets are not massive enough, and there are not enough of them.

Because there are so few interstellar planets, the odds against our Solar System encountering one are even longer than the odds against it encountering another star. If it did, then perhaps the stray planet could be captured by the Solar System. It might well be in an eccentric orbit and inclined at a large angle to the plane of the ecliptic in which all the other planets orbit. Fogg points out that this description matches some descriptions of the hypothetical Planet X.

Planet X is the unknown tenth planet of the Solar System. Its existence is hypothesized in order to explain discrepancies in the orbits of the outer planets, like Uranus and Neptune, whose orbits are not properly calculable, even after Pluto has been taken into account. But searches for Planet X in the ecliptic have been unsuccessful and it is hypothesized that Planet X must be in an unsurveyed part of the sky, in an inclined orbit.

R. A. Harrington and T. C. van Flandern of the US Naval Observatory have given a role for Planet X in the evolution of the Solar System. They propose that Planet X is not much more than Earth-sized and that it encountered Neptune at a time when it had a large satellite; the satellite was broken free from Neptune and became Pluto, according to this hypothesis.

Another role for Planet X would be to disturb the Oort Cloud. The Oort Cloud is the name for the hypothetical region where comets are stored, in the far reaches of the Solar System. Once in a while, according to Jan Oort, something happens to disturb the Oort Cloud and a comet breaks free, falling towards the Sun. Usually, the comet loops once round the Sun, and disappears back out into the distant far regions; only if the comet encounters a planet does its orbit alter, becoming periodic. If something, like a planet, strongly distrubs the Oort Cloud it may be that comets fall out of it in showers; this would happen periodically, as the planet repeatedly disturbed the Oort Cloud. This is a role for Planet X, according to J. J. Matese and D. P. Whartmire. The Louisiana astronomers attribute the periodic mass extinctions of life and the periodicity in cratering on Earth to the effects of the showers of comets on the Earth; the period, thirty million years, would correspond to the period of Planet X, captured into a highly elliptical orbit from interstellar space.

We have followed interstellar planets from their origin near to a star like our Sun, to their terrifying ejection into space by an encounter with another star or as a result of their sun exploding as a supernova. Perhaps the luckier of these worlds may be recaptured ' y another sun, so that they again see a feeble cycle of dim day to elieve the dark night. What a sign of hope this would be, after ιeons coursing, lonely, in the blackness of interstellar space!

Detecting Solar Radio Waves

RON HAM

Most astronomers already know that the Sun can be a powerful emitter of radio waves when sunspots are present; in fact, it is the detection of these waves with sensitive radio equipment on Earth which warns us when the Sun is active and that streams of charged particles may be heading toward our planet. The radio waves take a mere 8.6 minutes to cross the 93 million-mile (150-million km) path between Sun and Earth, but another 20 to 40 hours may elapse before the solar particles, which left at the same time, reach the Earth's orbital path.

Remember that the Earth spins on its own axis whilst travelling in orbit around the Sun and, if the timing is right and these particles collide with the Earth's polar atmosphere then an aurora is likely to manifest at either or both poles, and/or the normal state of the Earth's ionosphere is disturbed with a consequential upset to terrestrial communications. Although an aurora is a display of natural beauty, it is also a region of temporary ionization which, while it lasts, has a strange effect on VHF radio signals. For instance, when reflected from an aurora, the normal clean and sharp note of a Morse-coded signal becomes a low pitched 'rasp', some voice transmissions sound like a ghostly whisper and television pictures can take on additional distorted images. On the other hand, disturbance to the upper layers of the ionosphere can cause blackouts, echoing and fading on those transmissions that are designed to use its natural wave-bending and reflecting properties to reach their intended destination, often in the opposite hemisphere. Therefore, under these conditions, a wireless operator can hear the result of solar activity and, in the case of aurora, know of its presence even if clouded skies, a bright moon, or the hours of daylight prevent it from being seen.

A Simple Radio Telescope

During the mid-1960s it became obvious that my writings about the propagation of radio waves would carry more impact if I knew when the Sun was active, so, early in 1968, I built a simple radio telescope which observed the midday Sun, almost daily, for the

Figure 1. The 136-MHz antenna.

following sixteen years. Before building this instrument I had established that the radio waves from solar storms were detectable between 100 and 200 MHz, but, here on Earth, this part of the radio-frequency spectrum is extensively used by the public services and the number of quiet spots ideal for solar work were then and even more so now, extremely limited. After a period of tests I settled for the lower end of the 136/137-MHz satellite band where, at my location, other users were seldom heard. This is a vital decision because, for maximum efficiency, the antenna and receiver must be designed around it.

155

The Antenna

At a convenient site, facing almost due south, three 6-ft lengths of 9-in. × 3-in. timber were dug into the ground to support a 10-ft × 6-ft, wood-framed, wire-mesh reflector which carried four 4-element yagi antennæ, each cut to the observational frequency and electrically harnessed together (Figure 1, above). The frame was hinged on its bottom rail and back-supported, so that it could be adjusted about five times per year to keep the Sun within its vertical beam-width and the west–east rotation of the Earth carried the antenna, horizontally, through the path of the midday Sun. Although wildly exaggerated, Figure 2 shows that by this method the antenna **A** would first 'see' the Sun around position 3, observe its output at 4 and 5, lose it between 5 and 6 and wait about 21 hours before starting again, next day, at position 3. This system is commonly known as 'Earth Drift'.

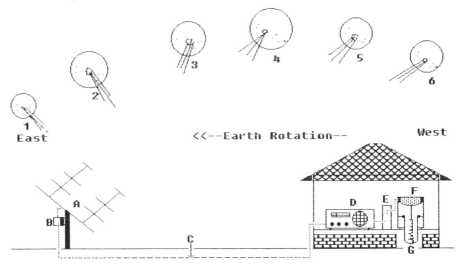

Figure 2. Shows the system commonly known as 'Earth Drift'.

Receiver

This radio telescope was based on the superhetrodyne principle and occupied a fair bit of space, because the three-stage radio frequency section (Figure 2 **B**) of the superhetrodyne was in the garden behind the reflector and the rest of the receiver, **D**, d.c.

amplifier, **E**, and pen-recorder, **F**, was situated in our house, some 150 feet away. The output from **B** was linked to the input of **D** by a good quality coaxial cable, **C**, laid underground in a protective plastic hose-pipe. The installation was operated by a timer which started the equipment at 11.30, when the Sun entered the antenna beam-width and turned it off again at 14.30 as it left. The observational results were clear because the antenna was situated in an interference-free location and the daily recording was spread along 7.5 feet of paper chart, **G**. Although costly, this high chart-speed of 0.5 inch per minute made it easy to distinguish between the background noise, produced by the receiver and any activity on the Sun. Incidentally, if you listen to the receiver noise, without the antenna connected, it sounds like a gentle 'hiss' and the solar noise, a positive 'whooOOoosh', therefore it is advisable to have a loudspeaker connected during the observation so that thunder static or man-made interference can be quickly identified and the chart marked accordingly.

How It Works

Now refer again to Figure 2, and follow an observation through the system from the antenna to a trace on the recording chart. Any solar radio waves at 136 MHz arriving at the antenna, **A**, were fed by a very short feeder to the first stage of the radio frequency section, **B**, where they were amplified and passed to a 'mixer' in **B**. This mixer also received a signal at 110 MHz generated inside **B** by a crystal controlled oscillator. The difference or intermediate frequency (136–110 = 26 MHz) then carried the incoming waves, at 26 MHz, along the underground cable, **C**, to be amplified again and processed by a good quality communications receiver, **D**. The line was finely adjusted to 26 MHz by the main tuning dial on **D** and any incoming solar waves at the antenna were then heard through the speaker attached to **D**. The voltage at the detector stage of **D** was too small to move the pen-arm on a recorder, so a d.c. amplifier, **E**, was built to boost this voltage and connected between the detector stage and the pen movement coil of the recorder. I used a separately powered 741 operational amplifier chip for this purpose, but please **do not** *attempt this work unless you really know what you are doing.* When all was working the receiver 'hiss' drew a steady line along the lower edge of the chart and a 'whoosh' from the loud-speaker, caused by a solar burst, corresponded to a swing of the pen-arm making its trace on the chart.

Results

Following a series of 'on-air' tests in May 1968, my observations began officially on June 1 and I soon found two distinct features of the solar radio output associated with sunspot activity. First, the isolated burst, Figure 3, with a life of between 0.5 and 10 minutes, and second, the continuous noise storm which can last for several days. One such storm is being traced on the right-hand side of the pen-recorder in Figure 4. Many times I noted that a major noise storm was preceded by an increasing number of individual bursts.

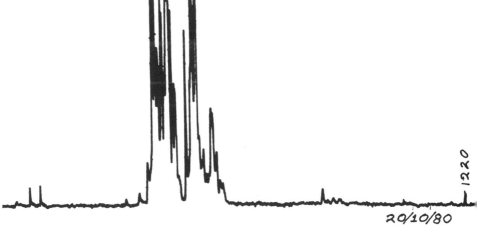

20/10/80

1220

Figure 3. Individual burst of solar radio noise.

The Sun rotates in approximately 27 days, and if a large sunspot appears on the east limb of the solar disk it will only be visible from Earth for approximately 13 days, but if this spot has a very long life it will re-appear again some 14 days later and, if still active, can create the similar terrestrial problems as it did the first time around.

A Memorable Event

Throughout the life of my telescope I recorded a multitude of differing solar bursts and a good number of noise storms, but one in particular really showed that consistent observation and co-operation with allied sciences was all worth while. The Sun had been relatively quiet during the last week in July 1972 and when my telescopes switched on at 11.30 on August 1, a local thunderstorm was in progress and each flash of lightning drew an unwanted

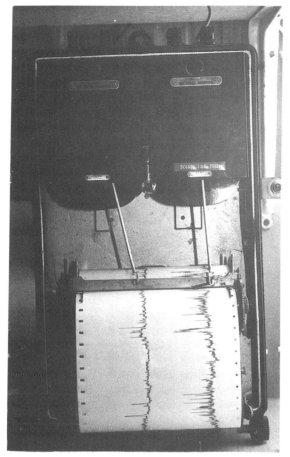

Figure 4. Solar noise storm being recorded on right side of pen-chart.

vertical spike on the recording chart (see Figure 5, bottom left). Suddenly, at 11.46, a massive solar burst occurred and, for most of its eight-minute life, it overpowered the local static. Because of the interference to radio equipment caused by lightning discharges, thunderstorms are a pest during any radio observation. However, this time it proved scientifically valuable and demonstrated the enormous power of a solar burst. Remember, on this occasion, the static was being generated within a mere mile of my location, but the solar radio waves that obliterated its effect on my chart originated 93-million miles away and had taken 8.3 minutes to reach my antenna. It was obvious from this that something big had started on the Sun, and very soon I was told by astronomical colleagues about the appearance of a large sunspot group (see Figure 5, top left). As the Sun entered my antenna beam each day the continuous radio noise was stronger, reaching its peak on August 4 as the sunspot-group was around central meridian (Figure 5, days 2 to 5) and although the level began to decline on August 6, this group remained active for a few more days. *The dotted line represents the normal receiver noise level.*

From sunrise on August 4, the solar radio waves were so strong that other equipment observing meteors, with a horizontal antenna, on 70.31 MHz, could not be used for several hours until the Sun was out of its beam much higher in the sky. When the telescope switched-on at 11.30 the solar waves sent the pen-arm against its upper stop where it remained throughout the observation.

By midnight on August 4 this storm had been raging on the Sun for three days, and one can imagine the orbiting Earth running the

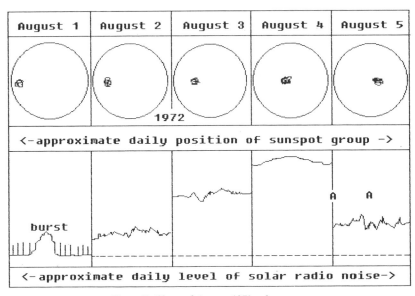

Figure 5. Chart of August 1972 solar storm.

gauntlet of the ejected solar particles. About 23.30, the sky was dark and clear and my attention was drawn to an arc of white light on the northern horizon about 20° wide and 5° high in the centre. As time passed several rays of light, a few degrees apart, reached over my zenith and as they moved across the sky from west to east their delicate shading changed from red to green to light blue. It is very rare to see an aurora from Sussex and I enjoyed every minute of it, especially the climax around 02.00 on August 5 when the bright stars of Ursa Major were shining through the pink auroral glow, which, by that time, had become the backdrop behind the beams and periodic patches of bright light. I later learnt that this aurora had an umbrella effect on VHF radio signals, and had the sky been overcast no one, except these radio operators, would have known of its existence. The solar storm continued that day, and another aurora manifested at 15.00 hours, but this time it was broad daylight and it was only detected by the strange effect it had on terrestrial radio signals. Before the advent of radio-communications untold numbers of auroræ must have gone by unrecorded because their beauty was hidden to the visual observers by overcast skies or they manifested during the hours of daylight.

Sunspot Cycle

Obviously, during the period of sunspot maximum, be it anything between ten and twelve years, major events occur more frequently, but I learnt not to be fooled by the numbers of visible spots, because it only takes a single group, or perhaps a couple of small spots, to cause an episode like the one illustrated in Figure 5. Therefore, anyone making radio observations of the Sun should keep this in mind and be very patient, especially through the years of sunspot minimum.

Photographing the Aurora

DAVID GAVINE

As darkness fell on the evening of March 13, 1989 amateur astronomers all over Britain, preparing for the night's work on a clear sky, became aware of vague green patches of light moving all over the heavens. An immense display of the Aurora Borealis, or Northern Lights, was in progress, following a massive solar flare. It was possibly the best aurora for thirty years, visible as far south as Spain and the Caribbean. Observers on the south coast of England were astonished to see the auroral arcs and bands in the *southern* part of the sky. Neil Bone and a group in Lincolnshire, favoured with a clear sky all night, saw several 'peaks' in which the red and green auroral rays, like long searchlight beams, converged from all parts of the sky to form a 'corona' in the magnetic zenith, that is, the part of the sky a few degrees south-east of the true zenith, to which the south pole of a magnetic dip-needle points. Unfortunately their colleagues in Edinburgh, having just witnessed an awesome blaze of yellow-green auroral light which lit up the ground and cast shadows, lost the display by 23.00 hours when cloud moved in from the west, and were forced to console themselves in a nearby tavern.

For many people in England it was their first big aurora, and an opportunity to try to photograph it, but with little warning or time to prepare, all kinds of cameras and films were hopefully pointed at the sky. Although some results were very good indeed, most found their pictures over- or under-exposed, or blurred, showing just a featureless sheet of green or red light. Photography enhances the colours of the aurora but cannot possibly capture the movements or pulsations of a great display, when sharp rays ripple horizontally, or bright blobs of light flare up and die, or the rayed band sways like a curtain in a breeze, or waves of light rush from horizon to zenith, apparently lighting up the rays and patches like leaping flames, a type of activity known as 'flaming'. This is the problem – trying to match a fast film and a big lens aperture with the *shortest* possible exposure to catch the shapes of the auroral forms. Auroral photography is an art which requires the right kind of equipment and attitude for good results. Any film or camera will not do (the 'instamatic' type is no use); it must have a 'B' setting for time exposures from a few seconds to several minutes, as big an aperture as possible (at least f2.8), and be mounted on a sturdy tripod or stand with a socket head to enable the camera to point overhead. A cable release is essential, a lens hood desirable.

A standard 50-mm lens will suffice for ray-bundles, the corona or a bright section of an arc, while a wide-angle lens of 28 to 35 mm is useful for taking in almost the whole of an average display. A 16-mm Sigma or equivalent all-sky lens operating at f2.8 can cover most of the celestial vault on a single frame and gives a spectacular view of an extensive or coronal display, especially when pointed at the zenith.

A faint aurora usually looks white, like the Milky Way, because its colours are not bright enough to stimulate the cones of the retina, but photography of such an aurora usually reveals a pale green or red. These colours become apparent to the naked eye in a bright display, but individuals differ in their ability to see faint reds. The two strongest lines of the auroral spectrum are produced by oxygen atoms and molecules excited by energetic electrons from the solar wind bombarding the upper atmosphere at heights between about 100 and 500 km. The blood-red emission at 630 nm in the upper parts of the aurora spread terror in he Middle Ages with reports of battles and conflagrations in the sky: the shorter rays and arcs at lower levels emit an apple-green line at 557.7 nm. During intense activity there is sometimes a reddish lower border on a sheet of auroral light, as very high-energy electrons penetrate deeper into the atmosphere and excite nitrogen. If an aurora occurs soon after sunset or before dawn the long-ray tops sometimes appear white or blue – they are actually in sunlight.

A number of colour films suitable for auroral photography are easily obtained, for slides or prints. Such a film should be fast without coarse grain and be responsive to the red and green auroral emissions. Experienced observers in the BAA Aurora Section favour Kodak Ektachrome 400 and 200 (which can be uprated), Fujichrome 400, and for prints Fujicolor 400. The response to the 630-nm red is slightly different in Ekta and Fuji, the latter renders it more purple or violet. Kodacolor Gold 400 is also good, and in the late 1970s Agfa CT21 was a popular auroral film. Although 'slow' at ISO 100, it is very responsive to the auroral green and red, to which it gives a purplish cast. The results are highly 'technicolor' and striking, but not very accurate. Faster but more expensive films at ISO 1600 and 3200 are worth trying, because of the shorter ex-

Figure 1. The above is reproduced from a colour slide. It shows the aurora as an active, rayed band, on April 29–30, 1990 0045 UT, looking north-east to Andromeda; Kodak Ektachrome 400, 50-mm lens f1.4, 8 seconds. (D. Gavine, Edinburgh)

posures they allow. However, some emulsions are very sensitive to the sodium-lit haze which pervades most of our country, and pictures exposed for a few minutes can be ruined by a yellow cast. In fact, the observer should get as far away as possible from all light pollution. It is a safe practice, when handing in auroral, or indeed

Figure 2. A colour print of the above picture is reproduced on the cover. It shows the aurora as a quiet, sharp-rayed band, on April 25–26, 1989 0100 UT, looking west to Gemini; Kodak Ektachrome 400, 28-mm lens f2.8, 20 seconds. (D. Gavine, Edinburgh)

any other astronomical results, for commercial processing, to ask that the film should not be cut, because it is difficult for the processor to see where each frame begins and ends. Cut and mount them yourself, using a viewer.

Exposure times for photographing the aurora depend on the film speed, lens aperture, brightness and activity of the display: it is not really possible to give hard and fast rules. Duncan Waldron, an experienced photographer, suggests 20 seconds at f2 with 400 film as a fair average which will give at least a result. The great display of March 1989 was actually a difficult one to photograph because it moved around the sky and pulsated too quickly. The quiet, sharp-rayed aurora of April 25–26, 1989 was much better. If the aurora begins, as it often does, as a quiet 'homogeneous arc', a low, pale green or white bow-shaped feature above the horizon and centred around magnetic north, use a wide-angle lens at f2.8 and give it about 30 seconds or 2–3 minutes if faint. The indicator that the aurora *may* become active is a narrowing and brightening of the arc. This stage is worth a photograph, say at 20–30 seconds, and it may reveal structures not visible to the naked eye, such as a second faint arc or faint red rays above it. Suddenly, the arc may develop bright patches of light at intervals along its length which break up into folds or ray bundles. Choose an active part of the arc, use a 50-mm lens at f2 or faster, and wait patiently! These features have a habit of breaking suddenly into sharp, bright rays. Open the shutter when they are sharp and clear, close it when they begin to drift horizontally or become diffuse. If the rays are very bright, 5 seconds should catch them, otherwise try 10 or 15. As the display progresses the rays become long and may show red tops. Again, try the 50-mm and wide-angle lens, use exposures of 10 to 30 seconds and, rather than waste film by photographing indiscriminately, always wait for the auroral forms to sharpen. Coronæ should be photographed at f1.4 if available, and at short exposures such as 5 to 10 seconds, because rays and patches pulsate about them and tend to blur. All-sky pictures at f2.8 require longer exposures, 30 seconds or more, but beware of artificial lights lurking round the perimeter, or of moving ground-lit cloud patches. Always keep a record of the times, apertures and exposures of your photography, then you will learn by experience.

People often ask 'Where should I go to be sure of seeing the aurora?' or 'When can I see it?' It is all too easy to say 'go as far north as possible' but there are limitations. Most auroræ are seen in a narrow zone running from the northern tip of Norway to Iceland, Hudson Bay, and Alaska. Two intrepid BAA veterans, Dr Alastair Simmons and Peter White, regularly visit Arctic Norway and Spitzbergen in search of the really spectacular auroræ, which move and flicker much more rapidly than those seen in Scotland, but which are brilliant enough to be photographed in one second. Arctic conditions, however, play havoc with batteries and camera mechanisms, and cause film to become brittle and crack, to say nothing of the effects of the severe cold on the observer. Geomagnetic disturbances tend to follow the 11-year solar cycle, but the auroral 'peaks' occur about one or two years before and after the sunspot maximum which is often itself fairly quiet. Thus 1978–79 and 1981–2 were active years, as was 1989. On this basis, 1992 promises to be a fair 'aurora' year. But auroræ are not exclusive to these times. Great solar disturbances can occur at any time, and give events such as the huge display of February 8–9 1986, near solar minimum, for which, unfortunately, the northern half of Britain was clouded out. Then there are the quiet 'coronal hole' auroræ often seen in northern Scotland away from the solar maximum. In any one year there tend to be more auroræ around the equinoxes. Don't go to the very far north in summer in search of auroræ, the sky doesn't get dark enough, and in winter the weather can be hazardous. The best locations in the world are probably North Dakota, Minnesota, Manitoba, Alberta, and Saskatchewan, where the auroral maximum zone, an oval round the geomagnetic pole, comes furthest south in latitude. Fine photographs from North America frequently appear in *Sky and Telescope* and *Astronomy*. In Britain, based on the statistics of the BAA Aurora Section, the skies along the Moray Firth coast from Dornoch and the Black Isle to Fraserburgh appear to be the best, comparatively cloud-free and with little urban light pollution. Strathmore, north of Dundee, and the island of Tiree have a good reputation, but Shetland suffers a great deal of cloud.

In midsummer the aurora observers often turn to that other intriguing phenomenon, the tenuous ice-crystal Noctilucent Clouds, which appear in the Mesosphere at around 83 km where the temperature can get down to $-160°$ C. They are seen from about the end of May to early August, shining in the middle of the night with a pearly-white or electric-blue radiance in the twilight arch, illuminated by the Sun which is between about 6° and 16° below the horizon, when all the tropospheric clouds are in darkness. The twisted filaments and fine herring-bone patterns of the NLC make fine pictures, very easy to take because the movement of the formations is so slow compared to the aurora. Almost any colour film can be used, even slow film such as ISO 64 or 100, and Kodacolor Gold 200 is about the best, it has given superb results over the last few years in the hands of Holger Andersen and J. Ø. Olesen of Denmark. Striking black and white photographs can also be taken on film with good contrast such as FP4 at ISO 125. As a general rule 5 to 10 seconds at f2.8 on ISO 200 should capture the NLC, bracketing half a stop either way, and if you intend to submit your work to the scrutiny of researchers take pictures exactly on the hour, quarter past, and so on, so that the results of others can be compared. Noctilucent Cloud, which current research suspects as being on the increase, is often regarded as another 'Scottish' phenomenon, but in recent years it has been well seen at Cambridge and observed as far south as Chichester and Exeter, even photographed well from London by David Frydman. Observers in eastern England actually have the advantage over the Scots for NLC photography, having less cloud and better contrast in a darker midsummer sky. Those in the far north in summer also have the terrifying Highland Midge to put up with – it becomes more bloodthirsty at dusk.

To see and photograph Noctilucent Clouds, therefore, there is no need for the Briton to travel too far from home. But, in the experience of those who journey far to look for the Aurora, it is possible to spend weeks in an 'ideal' location and see nothing at all. So any spring or autumn holidays in the north of Scotland should be planned with more than the aurora in mind, especially if the family is taken along. A final word – don't forget to avoid the Full Moon.

References

Bone, N. 'The Great Auroral Storm of 1989 March 13–14', *Astronomy Now* vol. 3 no. 6 June 1989, 22–9, Intra Press, London.
—— *The Aurora*, Ellis Horwood, Chichester (forthcoming).
Simmons, D. A. R. 'An Introduction to the Aurora', *Weather* vol. 40 no. 5 May 1985, 147–55, Royal Meteorological Society, Bracknell.
Anon. 'Noctilucent Clouds', *Weather* vol. 41 no. 8 August 1986, 259–63.
Gadsden, M. and Schröder, W. *Noctilucent Clouds*, Springer-Verlag, Berlin/Heidelberg 1989.

Active and Adaptive Optics:

THE ULTIMATE IN TELESCOPE DESIGN

RON MADDISON

Introduction

Many of those wonderful introductory science books that were written in the first decades of this century had titles like 'The Romance of Modern Astronomy', or even just simply 'The Heavens', and often described the subject as being the 'Queen of the Sciences'. This was obviously because the study of the stars is without question the oldest of the conventional sciences, and because it deals with things on the largest possible scale.

But although these aspects may well be true there remains a more fundamental property that makes Astronomy unique. That is its total reliance on remote, non-interactive observation.

All the basic methods and techniques of investigation and research that are presently in use have been developed from painstaking, small-scale, laboratory work but the essential data upon which our understanding depends can only be derived by careful, passive observation.

There is no way that astronomy, or perhaps more accurately in the words of Sir James Jeans, the study of 'The Universe Around Us', can be regarded as an experimental science. We can hypothesize and theorize, and we can predict what should happen under changed conditions but the unimaginable scales of time and distance that are involved prevent us from doing any experimental work to check our expectations. All that we can do is to look around us and see if our guesses are being fulfilled in some other part of the sky.

Great care has to be exercised in gathering and interpreting the very feeble radiation that comes to us over these enormous distances, because it carries all the basic information that we need.

It is only comparatively recently that it has been realized just how vast the distances are, and, in consequence, just how faint the radiation is. It is the drive to collect this feeble radiation that has revolutionized the telescopes and detectors that we find in several of the world's largest observatories. What is the prognosis for the next few years? Have we now reached the ultimate attainable in telescope development?

In order to answer these questions we need to look at how traditional telescopes work and then to evaluate the improvements made in the last few years as a direct result of the electronic revolution.

Magnification

The first telescopes were developed as toys at the end of the sixteenth century and were called 'Magic Tubes'. Their use was purely terrestrial, and they were designed to give a pleasing magnification of some distant object. The advantages of that are obvious, and there were soon many applications that exploited this property.

High magnification can be achieved by making the front lens of much longer focal length than the eye lens; in fact, the numerical value of the magnification is the ratio of these two focal lengths.

Galileo found that a negative, or diverging, lens used as an eyepiece gives an image that is the right way up although the field of view is quite small. A much larger field of view can be obtained with a positive eye lens but the final image is upside down. This may be a nuisance in a terrestrial telescope, but is of no real disadvantage for astronomical purposes. (The addition of a further positive lens lengthens the telescope, but restores the upright image.)

It was soon realized that the useful magnification that can be obtained with any telescope is limited not only by the qualities of the optical surfaces and materials that are used, but also by the nature of light itself. There is also the very obvious influence of the atmosphere through which we have to look.

A further factor that separates telescope targets into two categories is the effect of scale. The extremely large distances to such objects as individual stars makes magnification virtually irrelevant. No amount of magnification is sufficient to reveal the surface structure of a star directly, because its angular size is far too small. It is only for targets which are already extensive and have large angular sizes that a modicum of magnification may be desirable. The planets, nebulæ, and galaxies are examples of such 'extended' sources that fall into this category.

All the stars, with just one obvious exception, are so far away that they appear as mathematically tiny points of light. The eyepiece impression of star images as small disks is an illusion due mainly to the effects of turbulence in the refractive gas layers that make up the

Earth's atmosphere. There is also the fundamentally important effect of diffraction of the light waves around the edges of the lenses or mirrors of the telescope that leads to the formation of a spurious disk image called the 'Airy Disk'.

You can remove the former effect of atmospheric distortion simply by putting your telescope into space, above the atmosphere. Unfortunately the latter effect of diffraction is an unavoidable consequence of the wave nature of light. It is a property which can be minimized, but not eliminated. As a result of these factors, there is a limit to the useful magnification of a telescope beyond which no further detail can be revealed.

For the types of extended sources mentioned above, there is a minimum magnification that is needed to bring fine detail into the range of resolution of the eye or whatever other detector is being used to record the image – otherwise this detail will not be recognized and will be lost.

Light Grasp

Another criterion that is exercised in designing telescopes is the light grasp or aperture of the front lens or mirror.

The intensity of light radiated from a luminous object falls as the radiation spreads away from the source into the sphere of space that surrounds it. Its brightness is determined by an inverse square law of distance, so that a source which is of the same intrinsic brightness as another, but which is twice as far away, appears to be a quarter of its brightness.

A telescope needs to have as large a diameter as possible to collect as much radiation as possible, so that the operator can see fainter objects at larger distances. This is why the instruments are called 'telescopes'. The word is constructed from roots that mean 'seeing over large distances'.

The usual aim of a telescope owner is to acquire the biggest aperture that he can afford, and so this is usually limited by the budget rather than by any intrinsic technical problem.

Nowadays the reduced cost and improved availability of very large mirrors has transformed the situation for most amateurs, although the very largest instruments can still only be built through international co-operation resulting in centralized equipment that is widely shared throughout the whole scientific community.

Resolving Power

The resolving power of a telescope is its ability to separate two very close images, and it is expressed as the angular separation that can just be resolved. It amounts to the ability to separate two adjacent Airy Disks which partially overlap but can just be distinguished. This is found to depend on two factors: (1) the wavelength of radiation being collected, and (2) the diameter of the light collecting surface.

Longer wavelengths lead to a lower resolution because the Airy Disks are larger than for shorter wavelengths. Shorter wavelengths give rise to higher resolution. The larger the diameter of the objective lens or mirror of the telescope, the larger is the theoretical maximum resolving power.

If the observer is using a wavelength close to the centre of the visible spectrum, then the angular separation (α seconds of arc) of two objects that are just resolvable is given approximately by the expression

$$\alpha'' = 5.7/A$$

where A is the aperture in inches.

Here you can see that the maximum resolution obtainable with a 12-inch reflector will be approximately half a second of arc. The 94-inch aperture Hubble Space Telescope should be capable of resolving 0.06 seconds of arc; the 200-inch Hale Telescope at Palomar could not do better than resolving 0.028 seconds of arc, and is always limited to worse than this by the Earth's atmosphere.

Mechanical and Engineering Limits

Cost has always been the most important factor in determining the design of large telescopes. The move away from refractors towards reflectors at the beginning of the present century was strongly influenced by the properties of the materials involved.

By this time mirrors and lenses were made of glass. Great efforts were put into developing pure homogeneous glasses for lenses that could be cast without having internal bubbles or striations. But it was already obvious that such lenses can only be supported from their edges, and it is quite impossible to prevent them from sagging under their own weight.

Glass is a super-cooled liquid rather than a crystalline solid, and it flows like pitch. Admittedly it has very high viscosity, but the delicately polished surface curves on a lens cannot be maintained as the telescope is swung about the sky. The glass is subjected to various distorting forces of gravity over long periods of time, and it inevitably warps.

On the other hand, mirrors can be supported from all points on their backs and, furthermore, the light is reflected from their top surfaces and does not have to penetrate any glass. There is therefore no dispersive effect and no selective absorption. The reflector collects light more effectively than the refractor.

When the 200-inch (5-m) Palomar giant was completed in 1948, its highly reflecting aluminium surface was held rigidly on a block of low expansion glass weighing 14.5 tons. A set of mechanical balance weights press on the back of the mirror to counteract the varying gravitational distortion, and the system works extremely well. The mounting, used to aim the telescope at different parts of the sky weighs over 500 tons and is canted over at the polar angle to make an equatorial mounting.

However, asymmetric mountings like this are an engineer's nightmare. The bearing loads are shared unequally and the lower bearing has to be a thrust bearing. A much simpler design which overcomes this difficulty is the altitude–azimuth system, which is symmetrical relative to the direction of the force of gravity.

One of the best examples of this type of mounting is the 20-inch (51-cm) reflector built by James Nasmyth near Manchester in about 1845. This instrument is now in the Science Museum, South Kensington, London (see Figure 1).

The design is particularly convenient for naked-eye observation because the eyepiece remains in a fixed position at one end of the altitude axis, irrespective of the pointing direction of the telescope. The observer rides on a circular base which is part of the azimuth bearing, and he looks horizontally into the eyepiece. With practice the operator can easily track a target across the sky by hand, in spite of having to adjust both axes simultaneously and independently. The main drawback at the time was that it was difficult to devise an automatic drive which would work smoothly enough for long-exposure photography, and so it was not immediately adopted. The well-established, single motor equatorial was not challenged even by the designers of the Palomar instrument.

The trigger that precipitated the revolution was the invention of the transistor, and the development of computer-controlled, stepping motor drives.

Auto-guiders were invented, which took the tracking of targets out of the hands of the observer. The detection and response mechanisms became so rapid that smooth guiding could be controlled by the motion of the target itself. It was no longer a problem to

Figure 1. Woodcut of James Nasmyth at the eyepiece of his altitude-azimuth telescope (circa 1845).

control two axes simultaneously and the altazimuth mounting was at last fully resurrected. Smooth tracking can now be assured by the large mass and inertia of the telescopes that removes any vibration, and the human eye and hand combination can no longer compete for accuracy.

The break with tradition was beautifully demonstrated by the Russians, with the building of the 237-inch (6-m) reflector at Zelunchuskaya in the Caucasus Mountains (see Figure 2). In this case the mounting is altazimuth and is much simpler than the conventional equatorial. The drive system works perfectly.

Unfortunately the mirror in this enormous instrument initially proved to be inadequate for the exacting demands that were placed on it, and several years have been spent in replacing and improving it. In consequence this vast telescope is only just realizing its potential power.

In those early days it seemed as if a single monolithic mirror much larger than the 200-inch might never be built. Gravitational distortions seemed uncontrollable, and sufficiently rigid structures were

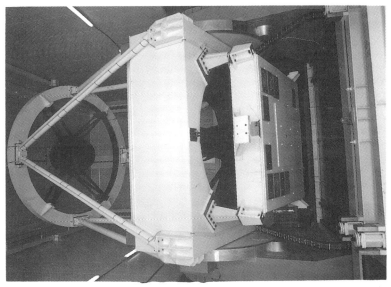

Figure 2. Photograph of the Russian 6-metre telescope showing altitude-azimuth system.

175

difficult to design and would have been prohibitively expensive to build.

Maybe the limit of size had eventually been reached?

Active Optics

One solution to the problem of the single monolithic mirror is the multi-mirror system. After all, radio astronomers use multi-dish interferometers, so why should the optical astronomer not do the same?

The answer was clearly because the wavelengths of light rays are so much smaller than radio wavelengths, and the necessary accuracy (better than one-tenth of a wavelength), seemed unattainable. The difficulties involved in supporting several mirrors and keeping them in constant register with one another while the whole structure is slewed across the sky were legion. The accuracy required is of the order of a millionth of an inch!

The major problem was how to monitor the attitude of each mirror with sufficient accuracy so that corrections could be made as soon as deviations occurred.

At this time enormous progress was being made in the comparatively new field of laser devices, and they were routinely being used to monitor such things as the changing distance of the Moon. Much smaller distances, on the laboratory scale, could be measured with unprecedented speed and precision, and it was 'just a small step' to apply the same technique to measuring the shape of a telescope mirror or frame. As a result, the Multi-Mirror Telescope (MMT) on Mt Hopkins in Arizona came into being (see Figure 3).

It was necessary to have a completely fresh approach to building large telescopes because it had at last been demonstrated that rigid structures are not essential. You can allow a certain amount of flexibility because you now have the ability to monitor it as it occurs. You can then immediately correct it by computer-driven mechanical and hydraulic jacks before it causes any image degradation.

This important step marked the transition from Passive Optics to Active Optics, and the new generation of super-telescopes had arrived.

The Multi-Mirror Telescope consists of six mirrors, each one 72 inches in diameter. The whole set is arranged in a hexagon so that the individual images can be combined on the central axis. The shape of the frame is monitored by means of a narrow laser beam

Figure 3. Mt Hopkins Multi-Mirror Telescope: front view showing mirrors.

reflecting between sets of corner mirrors placed throughout the system.

Each of the main mirrors can be individually set by means of three adjusting screws on its rear face, and final image coherence is achieved by small movements of the secondary mirrors. In this way image quality is preserved and the six component telescopes behave as one.

Experience gained with the MMT is now being used in the construction of a much larger sectioned, or 'tesselated' telescope. This is the Keck Telescope, which has a honeycomb-like structure made up of 36 individual mirrors giving a combined aperture of 390 inches (10 metres).

Once the problems of telescope frame stability and the over-coming of the inevitable gravitational distortions had been solved, it became possible to re-examine the difficulties of the single mono-lithic mirror.

Might it not be possible to save weight by making such mirrors very thin so that they become flexible? Perhaps they could then be controlled by bending, thus achieving a single active mirror which could be continuously maintained at its optimum shape? If this is possible then the well-tried ideas of the past will have been combined with the revolutionary ideas of distortable optics to give a system which behaves as if it is alive!

The first example of this technique was the successful completion of the New Technology Telescope (NTT) for the European Southern Observatory at La Silla in the Andes (see Figure 4). This has an aperture of 142 inches (3.6 m), is mounted in an altitude-azimuth cradle, and has auto-correcting optics of the 'active' type. The rear of the mirror is fixed to jacks which exert pressures where needed to give the best quality of image (see Figure 5).

Direct comparison of the first results obtained with those of another 3.6-m telescope of more conventional design on the same site, have already shown that very much better resolution can be

Figure 4. Photograph of the NTT at La Silla showing lightness of structure.

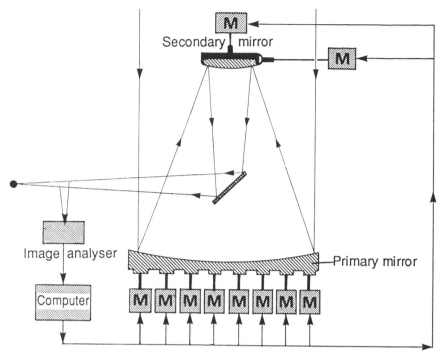

Figure 5. Diagram of jacks under the NTT mirror (from Astronomy Now, *Dec. 1988, p. 54).*

obtained. Once the telescope has been tuned to give its best performance the correction system is activated and maintains that setting. Experience shows that automatic corrections are made about once or twice a minute and optical checks on the figure of the mirror show that it is kept well within the desired specification. Furthermore, the structure is simpler and the whole telescope is a very much lighter piece of engineering than for any previous telescope.

This instrument has been so successful that it looks as if all future large telescopes will be built on the same pattern.

The New Giants

What are the implications of this for the next generation of post-Palomar giants?

There have already been significant additional advances in optical manufacture that have extended the size range of mirrors up to the region of around 315 inches (8 metres). It has always been the case that the most expensive and time consuming process in mirror making is grinding away large quantities of glass to produce the steep curves that are required. Recent advances in mirror casting at the Steward Observatory Laboratories of the University of Arizona have successfully exploited the well-known effects of centripetal force in rotating fluids.

A tank of molten glass slowly spinning causes the liquid to flow away from the axis of rotation and pile up against the outer wall of the tank. In practice the surface becomes almost parabolic in cross-section and the depth of the curve depends on how fast the tank spins.

The experimenters in Arizona have constructed a rotating furnace that is capable of producing thin mirror blanks that have almost the right parabolic shape and can cope with up to 8 metres in diameter.

Modern low expansion glasses, having thermal expansion coefficients of zero, can be cooled and annealed much quicker than, for example, the glass used in the 200-inch Palomar telescope, and it is now possible to start final working of the surface within just a few weeks of casting rather than the nine months or so required for the 200-inch.

Already there are several 8-m 'New Technology Telescopes' under construction and there is one 'Very Large Telescope' (a suitable piece of engineering understatement), which will consist of four such separate dishes operating both as a single instrument and as an interferometer. This will have the equivalent light grasp of a single mirror 630 inches (16 m) in diameter and it is to be built high in the Chilean Andes at a place called Cerro Paranal which is on the edge of the Atacama desert. This is perhaps the driest place on Earth, where the astronomical seeing is probably unequalled.

Adaptive Optics

'Seeing', or the optical quality of the atmosphere, is the remaining problem that prevents the attainment of the theoretical maximum resolution with any Earthbound telescope. A few independent research groups have been examining this problem, and there are encouraging signs that a new technique will help reduce

Adaptive Optics Test

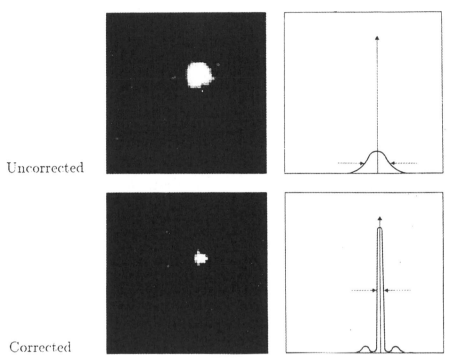

Uncorrected

Corrected

Figure 6. The effects of an adaptive system on image quality: (Top) *Uncorrected photo and light profile of the star Deneb using 50 seconds of exposure time.* (Bottom) *Same observation but with adaptive system switched on. (Reproduced courtesy of European Southern Observatory).*

the distortion significantly, although it is well recognized that it can never be totally eliminated.

The air above a telescope is always in motion. There are great bubbles of turbulence which refract the rays of starlight causing the twinkling effect that is so familiar to us. The flares and flashes seen through an eyepiece are quite rapid, and any integrating device such as a photographic film can only add them together and produce a sort of blurred average image.

By sampling the image electronically and quickly detecting its random jumping motion, it is possible to actuate thrusters on the back of a mirror so that the final image can be guided continuously on to one spot (see Figure 6).

But even for one of the new generation active optics telescopes it is quite impossible to superimpose such corrections on to the primary mirror. This is because that is too massive and could not possibly respond quickly enough. To cope you would need to correct the shape maybe as much as two hundred times a second. However, it has been shown to be possible to use a much smaller mirror further down the optical train, where the beam has converged, so that it is perhaps only a few inches across. Under these circumstances the mirror can be given a rapid response time by making it very light. It might be, for example, just a small fraction of a millimetre in thickness and about 10 cm in diameter.

Rapid corrections can then be made using piezo-electric thrusters and laboratory tests suggest that substantial improvements in image quality can be achieved. This technique is still in its very early stages but it has given rise to the new term 'adaptive optics' where the distorted image is massaged back into shape quicker than the atmospheric changes can distort it.

Space Telescopes

We all know of the unfortunate circumstances surrounding the testing of the Hubble Space Telescope, but its initial performance shows beyond doubt that the theoretical limit of resolution is within its grasp from an orbit high above the Earth's atmosphere. For bright objects the small amount of spherical aberration caused by the primary mirror can be ignored and the HST works extremely well (see Figure 7). The image degradation becomes apparent only for very faint objects, but since this affects a large part of the work of the telescope its presence is particularly frustrating. No doubt improvements will be made, but it will be some time before the very high costs involved will allow another such instrument to be launched.

For the next few years astronomers are bound to depend heavily on any improvements that can be made in terrestrial telescopes.

The Limits

The present position is that any desired optical curve can be manufactured and maintained using active components. We can therefore match the performance of the free-fall orbiting observatory in micro-gravity.

Figure 7. (Above) *Saturn as seen from above the Earth's atmosphere by the Hubble Space Telescope when its distance was 860 million miles.*
(Below) *Detail compares favourably with Voyager pictures taken at close fly-by at distances less than one million miles. This is also a Hubble Telescope picture.*

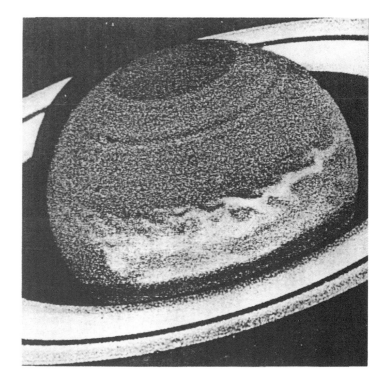

We can also minimize the worst effects of atmospheric turbulence, but even at the best sites there remains enough to prevent, for ever, the attainment of theoretical resolution. There is no way we can achieve from Earth the same quality of image that we can get from an observatory in empty space.

As far as image detection is concerned the present generation of Charge Coupled Devices (CCDs) can be used to count individual incoming photons, and the limit of detection is set by the brightness of the dark sky.

It is back-scattered light from cities and effects such as auroræ that set the limit for us, and it is unlikely that we will ever be able to reduce this interference.

On the face of it there seems little prospect of any further fundamental advance in large telescope performance. Let's hope we haven't blown our last chance of getting better instruments into space! The spherical aberration in the Hubble Telescope could so easily have been avoided – we may see the effects of this unfortunate error in the funding of planned space missions for many years to come.

Britain's First Professional Astronomer

GEORGE BIDDELL AIRY (1801–1892)

ALLAN CHAPMAN

January 2, 1992 marks the 100th anniversary of the death of G. B. Airy, who died after what was by any standards a remarkably long and eventful life. In an age when the death rate was much higher than it is today, Airy's forty-six-year term of office as Astronomer Royal was longer than most people could expect to live. Before becoming Astronomer Royal at the age of thirty-four, he had already achieved international distinction, while most of the ten years which remained to him after retirement at the age of eighty still contained scientific activity. Airy's status in the Victorian scientific world was formidable, and not least among his achievements was his development of astronomy as a profession out of which he earned a substantial livelihood.

We often hear it said that in the past, all astronomers were really amateurs. This statement has something to commend it, though it is not easy to translate the 'amateur' status of many seventeenth-, eighteenth- and nineteenth-century astronomers into modern terms. John Flamsteed, the first Astronomer Royal, was a professional in so far as he held an astronomical job, though his salary was quite inadequate in itself. Before receiving a modest Royal pension of £200 per annum in 1782 William Herschel made a very comfortable (and much larger) living as a musician, while the Victorian pioneer of the reflecting telescope, William Lassell, financed his activities from the fortune which he made as a Liverpool brewer. Even astronomical professors at Oxford and Cambridge were hardly 'professionals' in the modern sense, when their salaries were often less than those of bank clerks. The Lucasian Professorship of Mathematics at Cambridge paid a salary of £99 per annum in 1825, while Dickens's fictional *Nicholas Nickleby* earned £120 a year as a book-keeper in a London office.

What all of these men needed to do serious, high-level astronomical research was a substantial additional income which derived from a non-scientific source and which still gave them abundant leisure. John Flamsteed, the university professors and many of the Fellows of the Royal Society, for instance, were Anglican clergymen who enjoyed substantial ecclesiastical revenues with which to augment their minuscule or non-existent scientific incomes. Others were landed gentlemen, or else, like Airy's friend Richard Sheepshanks, came to inherit industrial or commercial fortunes which relieved them of all need to earn a living. Some were scientific aristocrats like the Earl of Macclesfield in the eighteenth century or the third and fourth Earls of Rosse in the nineteenth. Others, by a different direction, financed their astronomy from the proceeds of a successful profession, such as Herschel, Lassell or the German physician-cum-comet-hunter, Heinrich Olbers. Sir James South married an heiress in 1816 and promptly abandoned a successful surgical practice to build the finest private observatory in England on the proceeds of his wife's fortune. Thirty years later, William Rutter Dawes did a similar thing by marrying the widow of a rich Lancashire lawyer whose dowry made possible the commissioning of a sequence of large refractors with which he made an illustrious series of discoveries.

Some astronomers even combined several of these options: Sheepshanks, in addition to his private means, was also an Anglican clergyman, a barrister, and a Cambridge don, while Dr James Bradley, Astronomer Royal between 1742 and 1762, preceded his office with Holy Orders, an Oxford professorship and private means. When the friends of the Reverend Dr Nevil Maskelyne (who was also a Wiltshire squire) reminded the Prime Minister in 1810 that the Astronomer Royal had not received a pay rise in over forty years, the claim was dismissed on the grounds that Dr Maskelyne was a rich man and did not need any more money!

All of these men were 'amateurs' in the respect that while they devoted their creative lives to astronomy, only a small part, if any, of their income derived from it. But the whole social basis of astronomy was different from what it is today, because neither the state nor academic institutions saw it as a thing with which to burden the public purse. The duty of government was seen as consisting in little more than the provision of defence, law, and order. Science, art, music, and theatre were decidedly not government functions, and while they were seen as essential ornaments to a civilized nation they had none the less to be paid for by private individuals. Even the universities in 1800 saw themselves as *teaching* rather than *research* institutions, existing primarily to polish the minds of the nation's rising rulers and maintain religious orthodoxy. The pursuit of abstract investigation was not part of their function. Indeed, the only science in which the state was willing to invest was that which related directly to public well-being or defence, such as maintaining the Royal Observatory as a benefit to the Navy. The Oxford and Cambridge observatories likewise occupied themselves with 'useful' meridian astronomy, to justify their endowments.

In this essentially *laissez-faire* world, all intellectual scientific advance belonged to the private realm, making it essential for persons who wished to 'do research' in the modern sense to be 'gentlemen of means'. But George Biddell Airy was not a gentleman of means. His father was a relatively minor public official, and while there was some more money on his mother's side, he had no expectation of ever being independent. Airy always saw himself therefore as a 'working astronomer', who in the 1820s was faced with the problem of trying to make a living from a 'profession' in which there were no real jobs. It was hardly surprising, therefore, that even after a brilliant Cambridge career, and at the commencement of his international recognition, he considered re-training for the Bar as a way of making a secure livelihood. Fortunately, his Cambridge prospects improved dramatically, making this step unnecessary. But Airy's hard-nosed financial realism in a world of affluent astronomical gentlemen went a long way towards the creation of the *paid* profession of astronomer.

George Airy was born in Alnwick, Northumberland, on July 27 1801, where his father was stationed with the Excise Service. His real affinities, though, seem to have gone towards his maternal uncle Arthur Biddell, a farmer and land agent of Playford near Ipswich; a connection which was made easier after his father's posting to Essex after 1810. By his early teens, George came to spend increasingly long periods of time at the house of his uncle, who was a cultured man with an excellent library. Arthur Biddell, fascinated perhaps by his nephew's intellectual precocity, assisted with his education and he secured a place at Colchester Grammar School. Through his uncle, George also came to meet several prominent local persons and Cambridge dons, who advised in the securing of a place at Trinity College, Cambridge, when he was eighteen.

But Airy entered Trinity at the bottom. He was never a Gentleman Commoner, paid for by his uncle, but a lowly Sizar, or student-servant, who received a free education in return for the performance of menial duties. He soon won a college scholarship, however, and then set out systematically to win every prize for which he was eligible from his more dignified scholar's position. Airy's habits of systematic thinking, reading, and analysing made him an undergraduate phenomenon, and laid the foundation for that formidable capacity for disciplined work which was to stay with him for the remaining seventy-odd years of his active life. In his final examinations, he won marks that were twice as high as his closest runner-up (or Second Wrangler) and the highest marks ever awarded up to that time in the Mathematical Tripos. He was also an accomplished classical scholar with a love of Latin and Greek literature that was to remain with him to the end of his life.

Airy was elected to a short-term Junior Fellowship, or tutorship, of Trinity College, during which time he had the chance to show his colours as a teacher. Even as a student, in fact, he had offered private, paid tuition to older, richer, and less gifted contemporaries, and by the time that he was twenty-four years old he was making over £150 a year from teaching in Cambridge. He always had a long waiting-list of potential private pupils.

Yet while he made a good living in a college position where accommodation and meals were also thrown in as part of the job, it did not offer the future which he desired. For one thing, before Airy could convert his Junior to a Senior Fellowship of Trinity with full security of tenure, he would have to be ordained into the priesthood of the Church of England. All senior dons had to be in Holy Orders in 1825, even if they were mathematicians and astronomers, and Airy records that he possessed no religious vocation. Second, Senior Fellows had to remain bachelors, and as Airy had already fallen in love with, proposed to and been 'kept on ice' by a Miss Richarda Smith, whom he had met on a walking holiday in Derbyshire in 1824, he needed a job which enabled him to marry. This was why he considered giving up astronomy and becoming a barrister, though a fortunate circumstance arose in 1826 which allowed him both to stay in astronomy and eventually to marry: the vacancy of the Lucasian Professorship of Mathematics.

Unlike college fellows and tutors, university professors were at liberty to marry, and did not always need to be clergymen. As professorial endowments were often small, however, and lacked the perks and lucrative teaching fees of the college fellows, 'chairs' were often thought of as little more than part-time jobs. This is why Halley and Bradley had preceded the office of Astronomer Royal with university professorships, and why it was also handy to be a clergyman so that one could snap up additional revenues by holding absentee college livings. But as a married, un-ordained mathematical professor who was neither a barrister, physician, nor landed gentleman, Airy's financial future looked bleak.

Then in 1827 the much better endowed Plumian Professorship of Astronomy fell vacant, carrying as it did a salary of £300 per year, a house, and the Directorship of the newly established Cambridge University Observatory. Such an appointment was likely to attract powerful applications, though not necessarily from astronomers. Well-placed university dignitaries were not above gobbling up such positions, pocketing the revenues, and appointing some down-at-heel scholar to do the teaching at a fraction of the salary. Indeed, Airy's future friend, Adam Sedgwick, Senior Fellow of Trinity and subsequent Canon of Norwich Cathedral, had picked up the vacant Woodwardian Professorship of Geology in 1818 and freely admitted that he knew nothing about the subject. To his credit, however, Sedgwick took a crash course in geology at Oxford, delivered his first set of lectures, and was bitten by a scientific bug which turned him into one of the founding fathers of English geology.

But Airy tells us that as soon as he announced his intention of applying, the other candidates dropped out. Even so, university elections in those days resembled political campaigns, and Airy played his hand brilliantly. Having built up a corps of supporters, he had a 'manifesto' printed recommending an increase in the salary, and advocating Mr Airy as the best candidate. When the University played into his hands and offered him the Chair, he said that he would only accept it at the newly advocated salary, having no other perquisites to live on. The salary was then increased to £500, and in March 1828 he moved into the professorial residence and took charge of the Observatory. One cannot help but wonder what sort of career he would have had if he had followed through his earlier intention of becoming a barrister.

Once established as Director of the University Observatory, with a house and decent salary, Airy had become a 'professional' astronomer in every sense. He now set about turning the new Observatory into one of the finest and best run in Europe. He still lacked proper assistance at first, though Airy's *Diary* records that his spinster sister Elizabeth, who lived at the Observatory as part of the household, sometimes made 'unofficial' observations during her brother's absence.

During the seven years that Airy was Director of the Cambridge

Figure 1. G. B. Airy, from an engraving dated 1852. (Courtesy of Department of Astrophysics, University of Oxford)

Observatory he won the attention not only of the scientific community, but also of the government. Since 1834 Lord Auckland, the First Lord of the Admiralty, had been sounding him out about accepting the post of Astronomer Royal as soon as the current and ailing incumbent of that office, John Pond, died or retired. But Airy knew that it would be a daunting task, for the inefficiency of the current staff and sloppiness of work produced had reached the level of a national scandal.

After much hard bargaining, Airy accepted the office, but only after he had bludgeoned the government into giving him a pay rise of 25 per cent above that of his predecessor, arguing that, as a man of private means, Mr Pond was not wholly dependent on his £600 per annum salary, whereas as a poor man, he would be. The government had already tried to coax Professor Airy by offering a knighthood to him, although he turned it down on the grounds that honours without profits were without value. In 1835, when he eventually took up the office, it was agreed that Airy was to receive £800 per annum, plus a Civil List annual pension of £300 which had been offered to him but which he had settled on his wife. Eleven hundred pounds a year, plus an official residence, provided a very good living in 1835 and would have been equivalent to that of a comfortably-off lawyer or cathedral dignitary. It certainly made Airy the highest paid astronomer that England had ever known, and a 'professional' in every sense.

But if Airy, as a poor man, demanded a decent salary for his services, he was also willing to work full time to earn it. Most previous astronomers, including university professors and Astronomers Royal, had considered themselves to be masters of their own time, performing their tasks as and when they fancied. Airy, on the other hand, was willing to devote his entire working life, week in, week out, to his official duties. His working methods were quite at odds with those gentleman 'amateurs' who considered that education and social privilege made it unnecessary to be accountable to others on a daily basis. Indeed, Airy's rigour puzzled not only his fellow scientists, but also high-ranking civil servants (of whom he was one), admirals and politicians.

Airy's 'factory' approach to astronomy may have produced a great body of high-quality observations, but it was not easy to obtain highly educated staff, for in 1840, no man with a university education behind him expected to work a regular eight-hour day, six days per week, with only a month's holiday. Even the Admiralty, to which the Astronomer Royal was responsible, considered that his régime frightened men of talent away, though they were delighted at the quality of Royal Observatory work which the régime produced.

Airy's creation of a full-time professional astronomical institution left marks on the Royal Observatory which it carried down almost into our own time. Recently retired members of the RGO staff have told me that when they first went to work there in the 1940s, Airy's ghost still walked, as it were, as routines and procedures from the early Victorian era were still in operation sixty years after the old man's retirement in 1881. He certainly professionalized astronomy, but at the price of discouraging individual initiative and placing all stress upon the routine observation of meridian objects by a staff of 'drilled' technicians rather than science graduates. In the light of these conditions, therefore, it is interesting to see what type of men were willing to become professional assistant astronomers at Airy's Greenwich.

As soon as he took up office in 1835, Airy set about transforming the staff and creating a hierarchy which survived into the mid-twentieth century. His first act was to get rid of Pond's resident chief assistant, Thomas Taylor, a roguish ex-naval officer. By long practice, Taylor had come to run a private show of his own at Greenwich, taking bribes from London chronometer makers, giving the best assistant jobs in the Observatory to members of the Taylor family, working when he pleased and sometimes being drunk on duty. After feathering his nest for many years, the 63-year-old Taylor did not seem to mind being pensioned off and out of the reformed Royal Observatory.

Taylor was replaced by the Reverend Robert Main, a 27-year-old Cambridge mathematical don, glad to be appointed to a salary of £350 per year which, unlike his University position, enabled him to marry. Airy frankly described Main's future job as resembling a managing clerk in a large commercial enterprise, rather than that of an astronomer, for it would be his task to run the lower staff. But Main does not seem to have minded, for he stayed with Airy for twenty-five years.

Beneath this gentleman manager were two orders of non-graduates who did most of the observing and calculating. Airy consolidated the practice of employing bright youths in their late teens as 'computers' to be trained up to perform one specific type of calculation, or work beneath a senior assistant. It was from the ranks of these young men – of whom there were usually about a dozen in semi-casual employment at any one time – that the permanent 'warrant' assistants were appointed when a vacancy occurred. There were usually about six full-time warrant assistants, so called because they, like the 'professional' officers in the Navy (i.e. masters, surgeons, chaplains, and pursers), were appointed by

Royal Warrants from the Admiralty. All the permanent Greenwich staff members were warranted, including the Astronomer Royal himself.

In general, the men who became warrant assistants were the products of local private schools, who for one reason or another had never gone on to university. In an age when Oxford and Cambridge were only open to Anglicans, and before the University of London began to make an impact on education for non-Anglicans, few higher education openings were available to Catholics and Protestant dissenters. The majority of Airy's assistants fell into one or other of these categories: intelligent young Catholics and Methodists from middle-class backgrounds, excellently educated by their parents up to the age of sixteen or eighteen but with few opportunities thereafter. The Royal Observatory could give prospects to such men possessing mathematical skills to become the first professional 'assistant astronomers', and on average, their pay and working conditions were roughly the same as those of senior clerks in government offices. Working about forty hours per week, for salaries between £130 and £290 a year, they were solid members of the 'working' middle class, as opposed to the 'upper' middle class which was more financially self-determining. Even so, on £290 per annum in 1860, one could live in a comfortable villa in Blackheath, employ a couple of servants, enjoy a good holiday in Torquay and look forward to a handsome pension. By way of comparison, the Observatory gate porter and the labourer received £47 a year in the same period.

This was England's first full-time professional astronomical establishment, and we must now look at what it did in scientific terms. Most of the astronomical work performed at Airy's Greenwich was directed towards two ends: the ceaseless measurement of celestial co-ordinates in the meridian, and the observation of the Moon's position in relation to the fixed stars. This stellar, solar, and planetary positional astronomy was required for the compilation of accurate tables intended for use by the Admiralty, or any astronomers who chose to use them. At first, these observations were made with the large Mural Circle made by Troughton which Airy inherited with the Observatory in 1835, though between 1850 and 1950 they were taken with a superb Transit Circle of the Astronomer Royal's own design and built by Troughton's successor, William Simms.

The measurement of the Moon's position was concerned pri-

Figure 2. The Greenwich Transit Circle, 1850. Designed by Airy, the finely graduated 6-ft diameter meridian circle of this instrument denoted the Greenwich meridian down to our own time. Its telescope possessed an 8-in object glass. It was in regular use for 100 years and was a tribute to Airy's engineering skill.

marily with the improvement of navigation, for better lunar tables held the possibility of developing a simplified technique for finding a ship's longitude at sea. Airy's brilliance as an engineer and instrument designer was also evident in this area of investigation when, in 1847, he brought into use an Altazimuth instrument of his own design which made it possible to measure the Moon's position anywhere in the sky, rather than just in the meridian. But astrophysics, spectroscopy or the physical astronomy of the Sun, Moon, and stars were *not* a part of the Royal Observatory's functions, as far as Airy interpreted them. As far as he was concerned, it was not the duty of the public purse to finance abstract research, but to confine itself to strictly utilitarian tasks for the Navy and the state. This narrow concept of his duties undoubtedly removed the Royal Observatory from many of the most exciting enterprises of nineteenth-century astronomy, which led to its becoming relatively fossilized during the last decade of Airy's reign. Its one concession to the new branches of astronomy which were developing, however,

Figure 3. Airy's Altazimuth instrument, 1847. It was designed by Airy to observe the Moon outside the meridian on its 3-ft diameter circles. The whole instrument was devised to use as few bolted parts as possible, to minimize potential errors deriving from joints working loose.

was the establishment of a photographic department in 1874, to take photographs of the Sun for meteorological purposes.

Yet one area in which the Royal Observatory was the undisputed pioneer was time-finding and distribution. Airy was quick to recognize the potential of the electric telegraph in the early 1850s, and used telegraphically related systems of magnets and relays to operate self-recording instruments in the Observatory as well as transmitting daily time signals through the commercial telegraph network to the nation at large.

The performance of the tasks outlined above would never have been possible by an astronomer who only worked on an *ad hoc*

Figure 4. The Chronograph Room at the Royal Observatory. The large drums carried the paper sheets on to which the various instruments and clocks recorded their signals by means of electric impulses. In this way, Airy could reduce to a minimum any dependence upon an observer's skill, and make much of the recording automatic. Similar apparatus was also used, from the early 1850s onwards, to transmit GMT signals into the national telegraph network.

basis. It needed Airy's beloved 'daily attendance' and 'discipline' of technique into which the entire Observatory staff had been 'drilled'. In many respects, however, Airy had ceased to be an astronomer by 1840, and became instead an astronomical organizer and administrator. He was perhaps unique among the great astronomers in so far as he looked down upon the act of observing the heavens. Practical observation, he wrote in a letter of 1847, was a mindless task which could be performed by 'idiots' who had been thoroughly schooled in the use of high-grade instruments.

His opinion about practical astronomy seems odd in many respects, for all the great contemporary European and American astronomers whom he so admired, such as Bessel, Struve, Encke, and Bond, were masterful observers, whose astrometric work had led to the determination of stellar parallaxes and many other new discoveries. All of his predecessors at Greenwich had also been fine observers, and no one could say that they had not also been men of the highest intellectual attainment as well as practical skill. There were others, too, William Herschel had probably been the most spectacularly intellectual observer of all, while his son John Frederick and many Fellows of the Royal Astronomical Society were both accomplished observers *and* personal friends of Airy.

But it is my suspicion that Airy's personal disdain for practical observation stemmed from two sources. The first of these was his own poor eyesight. While still a young man at Cambridge he had applied his powers of optical analysis to his own eyesight, and detected pronounced astigmatism. In the early 1820s, this had been an untreatable disorder until Airy computed the curves and refractive indices of a pair of astigmatic spectacles which he then had made by a lens grinder. He wore astigmatic spectacles for the rest of his life, though one can understand how the disorder must have made practical observation an irksome task rather than a thing of wonder. What always fascinated Airy was the mathematical exactitude of the heavens rather than their shining glory.

The second thing which I suspect confirmed Airy in his low opinion of practical observation was the scandalous fiasco which threatened to tear the RAS apart in the 1830s and 1840s, and which involved him and his friend Richard Sheepshanks. This was the prolonged litigation and mud-slinging match which ensued when Sir James South refused to pay Messrs Troughton and Simms for mounting an 11¾-inch object glass for him on the grounds of defective design. Sir James, who was a rough-dealing, cantankerous and profoundly quarrelsome man, was also the finest double-star observer of the age, working from his magnificent private observatory in Kensington which he had paid for out of the proceeds of his wife's fortune.

Quite apart from the unsavoury details and crossfire of personal abuse which South hurled and which the equally bellicose Sheepshanks thoroughly enjoyed hurling back, I believe that Airy came to measure astronomical observation in terms of Sir James's personal character. The legally trained and quite uncharitable Reverend Mr Sheepshanks was perhaps not far from the truth when he accused Sir James – who was first, or Charter President of the RAS – of being an ignoramus who did not know the difference between a sine and a cosine! South, after all, was a *surgeon* by training and a gifted practical observer by instinct. But he was not one of the Cambridge mathematical élite which then controlled the RAS, and I suspect that Airy came to think that all practical observers were tarred with the same brush; difficult, ungentlemanly persons lacking the appropriate theoretical background. In short, not *real* astronomers at all, but glorified artisans. Amateurs in the worst sense.

Sir James South, for his own part, never let the public forget that 'the Astronomer Royal does not observe', and in 1847, almost twenty years after the first outbreak of hostilities, published a blistering critique of the latest Greenwich observations. From a statistical count, he found that of the 69,368 logged observations since 1835, Airy had only made 164 of them himself. It was in defence of these imputations that Airy stated that practical observation was an occupation for trained, obedient idiots rather than professional gentlemen. One can only wonder how his Continental friends must have reacted.

One thing which made it possible for Airy to be so disdainful about practical observation was the simple fact that the new Greenwich instruments which he had developed required relatively little skill to use correctly. Thanks to a revolution in precision engineering, self-recording electrical devices and rigid error-correction procedures, Airy's genius as an instrument designer reduced meridian astronomy to a dull routine procedure. While he may have been exaggerating when he said that trained idiots could use the new Greenwich instruments, it was none the less true that *carefulness* rather than *cleverness* was all that an operative needed to use them effectively. One might say that what Richard Arkwright did for the textile industry, Airy did for meridian astronomy.

Those intelligent skills in which colour estimates, complex micrometric measurements, and carefully planned courses of observations were needed – such as Bessel's parallax measures of 61 Cygni – were not needed at Airy's Greenwich because that type of research was not what the Admiralty funded the Royal Observatory to do. So while Airy may only have needed idiots or drudges at Greenwich, he could still admire the observational feats of his Continental colleagues, who pursued complex astrometric researches demanding theoretical backgrounds which were more highly developed than those of the loathed Sir James South.

As England's first full-time professional astronomer, one might expect that Airy made a number of major discoveries which changed the way in which people thought about the Universe. But he did not. Indeed, he set about the creation of a variety of yardsticks by which the natural world and Universe could be measured. Quantification lay at the heart of his work at Greenwich, and the provision of exact standards first to the Navy, then to the international scientific community. His determination of improved celestial tables mentioned above was one of these enterprises, but there were others. All of them, however, possessed a practical, utilitarian dimension.

One of the earliest tasks in this respect had been his attempts to obtain a more exact value for the density of the Earth, when in 1826, 1828, and 1854, he conducted experiments in deep mines. Precision pendulums were swung at the top and bottom of deep pits in Cornwall and on Tyneside, to measure the increase in gravitational acceleration upon a swinging pendulum at a given depth below the Earth's surface. Though these results were less decisive than he had hoped for, they displayed Airy's relentless concern with accuracy standards, along with a brilliance of experimental technique.

He also devoted great energy to the determination of the exact longitudes of places on the Earth's surface. This function clearly related to the Astronomer Royal's maritime responsibilities, but once again, he went on to obtain the very best standards using the latest technology. In 1844 he used chronometers to establish the exact location of the Greenwich Observatory from that at Altona in Denmark. This was to link Greenwich with the Pulkova–Altona geodetic baseline surveyed by Wilhelm Stuve, so that the observations of three major meridian observatories could be interchanged. Airy also went to great lengths to determine the precise longitude of Valentia in Ireland from Greenwich, representing as it did the most westerly accessible point in the British Isles. All of these longitudes were determined by multiple chronometer comparisons between Greenwich and the chosen location, to compare GMT with their local time.

With the introduction of electric telegraphy in the late 1840s, Airy realized that the almost instantaneous passage of electricity along a wire gave a far more reliable way of sending GMT for longitude purposes than transporting chronometers. The Royal Observatory was wired up to the Telegraph Office in Greenwich, and from there to Cambridge, Edinburgh, Liverpool, and other places of astronomical and maritime importance in the British Isles. In the early 1850s, the mechanically operated Time Ball on the roof of Flamsteed House was electrified, and signals sent to actuate similar balls in the major naval dockyards. Ship commanders in Deal, Plymouth, Spithead, and other places could now rate their chronometers with GMT before setting out to sea. With the completion of a submarine wire to France in 1851 signals were exchanged with Paris and the Continental observatories, and after the laying of the first Atlantic cable in 1858, with the United States, and with Australia after 1872.

By relating the longitudes of places to Greenwich, Airy went a long way towards the creation of the Greenwich Prime Meridian of the world, and when the international Washington conference assembled in 1884 to decide upon a 'Longitude Zero', Greenwich seemed a natural choice.

Mention has already been made about Airy's work on astronomical instrument design, and more should be said of this, for he was a man of great talent as an engineer. Airy approached engineering from a scientific direction, analysing the stresses and strains of girders, wheels, and beams in terms of applied mathematics, and then designing structures of optimum efficiency. His designs for the lunar altazimuth instrument of 1847, and the great Transit Circle of 1850 – which defined the Greenwich meridian through a century of daily observation – were masterpieces of structural engineering from first principles.

Airy designed all manner of instruments, large and small, with the intention of making the Royal Observatory's daily observations as simple and precise as possible. Having the low opinion that he had of his assistants and of practical observations, he set out to replace the human sense of hand and eye judgement (so necessary in taking meridian transits) with the electric micrometer. Instead of

Figure 5. The Royal Observatory in Airy's time. Notice the famous Time Ball on the roof. This device was first installed by Pond in 1832 to signal time to ships lying below on the Thames, although Airy made its operation automatic after 1852 when an electric signal from the main regulator clock triggered it. Note also the meteorological instruments positioned on the roof. Airy's Greenwich became the first British institution to make and publish detailed weather records.

having to count a transiting star's contacts with the meridian wires and compare them with the beats of the regulator clock, the electric micrometer needed only simple guidance from the assistant. Electric contact points in the micrometer now sent signals which actuated a punch prick recorder on the same drum on to which the beats of the regulator clock were being registered. Right Ascensions were thereby timed automatically, and instead of split-second timings being made by the observer, to ascertain the error of the clock, one merely compared the line of automatic punch marks on the paper. By 1855, many electrical devices were in regular operation at Airy's professionalized Greenwich, not to mention the daily transmission of public time signals to railway stations and telegraph offices taking GMT to the nation.

Over the course of his long term of office, Airy also conducted numerous non-astronomical investigations, to assist the Navy or provide technical advice to Parliament relating to legislation lying before it. It was Airy, for instance, whose researches into terrestrial magnetism established the laws governing compass deviations in iron ships, and provided mathematical rules for their compensation. As Second Railway Gauge Commissioner in 1845–46, his engineering expertise was called upon to assist Parliament in deciding between the 7-ft and 4-ft 8½-inch gauges for future British railways. Modern railway historians have still to tap the enormous body of Gauge Commission documents preserved in the Airy papers in the Archives of the RGO for the light which they cast on the state of English railways in 1845. Indeed, Airy advised the government on a range of technical issues that extended from the design of London's sewers to safety at sea: over thirty non-astronomical Commissions and Inquiries in all.

Some of Airy's critics have pointed out that he could have devoted more of his personal time to astronomy had he not been so willing to get involved in these non-astronomical public duties. While it is true that he organized the Royal Observatory to run like clockwork under the day-to-day supervision of the Chief Assistant, his régime none the less had a stifling effect on the scientific creativity of the institution, with its excessive stress on routine functions which discouraged active thought on the part of the staff and deterred creative scientists from joining it.

Yet most people on hearing Airy's name today think of him as the astronomer who failed to locate Neptune, on the co-ordinates supplied to him by J. C. Adams. I have written elsewhere on Airy's role in this affair, and can only re-iterate that to blame him for the loss of the discovery is seriously to misunderstand many of the factors bearing on the situation. Suffice it to say that Adams did not produce a set of polished computations on the strength of which any astronomer could really have commenced a search in 1845. After Airy had written in reply to Adams's preliminary communication of results in October 1845 to request further information, Adams did not even bother to acknowledge the letter. In fact, Adams did not get around to writing another letter to Airy until September 1846, almost a year later, by which time the Berlin Observatory was on the verge of discovering Neptune on the strength of Le Verrier's co-ordinates.

When Adams first wrote to Airy in October 1845, he failed to point out the mathematical processes by which he had come to his result. As a thorough and painstaking scientist himself, Airy desired to know these processes before he was willing to give the matter his serious attention. It was in request of this information that Airy had written to Adams on November 5, 1845, and to which Adams failed to reply. Without them, Airy would have been acting on blind faith: on the strength of figures sent to him by an unknown young man who had only recently completed his undergraduate degree.

The years 1845–46 were probably the most hectic of Airy's entire career. For one thing, he was in the midst of his work as a Railway Gauge Commissioner. Secondly, the latter part of October 1845 (when Adams made two un-announced visits to Greenwich and failed to find the Astronomer Royal at home) was especially traumatic for everybody in the Airy household. Airy's 42-year-old wife was about to give birth to their ninth child; while on October 27, evidence was placed before the Astronomer Royal that one of his senior assistants (one of the Old Guard who had avoided being sacked with Thomas Taylor) had committed two ghastly crimes. William Richardson had fathered a child upon his own daughter, after which they poisoned the healthy infant with arsenic and buried it in the garden. When workmen came to lay a pipe in the garden some time later, the unregistered burial was made public and Richardson and his daughter were arrested. A real-life Victorian murder melodrama was being enacted on the Astronomer Royal's own doorstep.

The fact that Airy could find the time to write to Adams at all in the midst of such turmoil shows his energy and level-headedness. But to expect him to chase up an obscure Cambridge graduate student pursuing what was probably a worthless mathematical speculation anyway was asking for perfection. Adams's long silence can only have confirmed Airy's suspicion that the query about detecting a planet beyond Uranus by the analysis of orbital perturbations had run out of steam even in its author's own mind. But even if Adams *had* provided exact co-ordinates and a convincing mathematical explanation, it was not the Astronomer Royal's job to look for it. Speculative, investigative astronomy was a job for a private or endowed university observatory. It was certainly not the job of the British taxpayer, and Airy was astonished to find that many of his fellow countrymen called him names when the discovery was made by the Germans on the strength of French figures.

During the course of Airy's long life he was the outstandingly successful English professional astronomer. But it is unfortunate that he acquired so few colleagues of comparable status. With the exception of the Directors of the Oxford and Cambridge University Observatories – who were also teachers as well as researchers – there were still hardly any men, and no women, who earned a decent living from astronomy. Official or paid astronomy remained limited in its scope to the fulfilment of routine tasks, not the pushing back of the frontiers of understanding, and front-line *research* astronomy still remained (in England) the province of the gentleman amateur down to the early twentieth century.

We must not forget that, while England was trail-blazing many new avenues of research by 1870, virtually all of them were funded out of private pockets. Airy's friend, the pioneer solar physicist, Warren de la Rue, enjoyed private money from the family's commercial interests. William Huggins, who started to identify the chemical make-up of the stars using the spectroscope from the late 1850s, was given his first observatory and livelihood by his father, a wealthy brewer. Joseph Norman Lockyer was also a rich man, whose magnificent private observatory passed through Exeter University and into the present control of the Sidmouth Astronomical Society.

The RAS continued to reflect this social composition in its Fellowship. It is true that the founding of London University in 1828, followed by Durham, Manchester, and the new Metropolitan Universities later in the century had weakened the Oxbridge stranglehold on English higher education, although astronomy did not figure especially either as a subject for teaching or an area of front-line research. It is true that the Scottish Universities had long since taught astronomy as a curricular subject, but with the exception of the Edinburgh Royal Observatory (which was similar in some of its functions to Greenwich) there were few large observatories. The Astronomer Royal for Scotland in the 1850s, moreover, received a salary that was lower than that of Airy's Chief Assistant. Ireland had observatories at Armagh and Dunsink, though it was still Greenwich-style meridian work, rather than the new pioneering branches of astrophysics that predominated. While Ireland was the home of the world's largest telescope at this time, which was used for fundamental, front-line deep-sky research, it was the private property of the Earls of Rosse and the scientific gentlemen whom they permitted to use it.

If Airy was Britain's first real professional astronomer, it is sad to note that at the time of his death a century ago he was still relatively alone in terms of position. It was in Germany, however, where substantial state funding was being poured into all branches of scientific research, that the real professional astronomers were to be found in 1892. And by 1910, America, with its Graduate Schools and millionaire observatory endowments, had become the home of the greatest number of men, and a growing handful of women, who earned their livings by probing the secrets of the Universe.

Yet in the nineteenth century, the high-level English amateur astronomer was a formidable and creative figure. It says much that George Biddell Airy not only learned everything that these men had to offer, and won their acclaim, but succeeded in contriving circumstances so that he could turn into a profession, and create as an international standard of excellence, that which was to them a form of leisure occupation.

Further reading

The author has been working for some time on a biography of G. B. Airy in the archives of the RGO and elsewhere, from which much of the above material has come. But see also:

G. B. Airy, *Autobiography*, ed. Wilfred Airy, Cambridge 1896.
A. J. Meadows, *Greenwich Observatory, Vol. 2; Recent History (1836–1975)*, London 1975.
Derek Howse, *Greenwich Observatory, Vol. 3; The Buildings and Instruments*, London 1975.
Michael Hoskin, 'Astronomers at War: South v. Sheepshanks', *Journal for the History of Astronomy*, xx, 1989 175–212.
A. Chapman, 'Science and the Public Good', *Science, Politics and the Public Good*, ed. A. N. Rupke, Macmillan, London 1988, 36–62.
——, 'Private Research and Public Duty; G. B. Airy and the Search for Neptune', *Journal for the History of Astronomy*, xix, 1988, 121–39.
J. G. Crowther, *Scientific Types*, London 1968, 359–92.
Patrick Moore, *The Planet Neptune*, Chichester 1988.

Some Interesting Variable Stars

JOHN ISLES

The following stars are of interest for many reasons. This year the list has been completely overhauled by including information from the most up-to-date sources, and several bright Mira stars have been added. The positions are given for epoch 2000. Of course, the periods and ranges of many variables are not constant from one cycle to another.

Star	R.A. h	m	Declination deg.	min.	Range	Type	Period days	Spectrum
R Andromedæ	00	24.0	+38	35	5.8–14.9	Mira	409	S
W Andromedæ	02	17.6	+44	18	6.7–14.6	Mira	396	S
U Antliæ	10	35.2	−39	34	5–6	Irregular	–	C
Theta Apodis	14	05.3	−76	48	5–7	Semi-regular	119	M
R Aquarii	23	43.8	−15	17	5.8–12.4	Symbiotic	387	M+Pec
T Aquarii	20	49.9	−05	09	7.2–14.2	Mira	202	M
R Aquilæ	19	06.4	+08	14	5.5–12.0	Mira	284	M
V Aquilæ	19	04.4	−05	41	6.6– 8.4	Semi-regular	353	C
Eta Aquilæ	19	52.5	+01	00	3.5– 4.4	Cepheid	7.2	F–G
U Aræ	17	53.6	−51	41	7.7–14.1	Mira	225	M
R Arietis	02	16.1	+25	03	7.4–13.7	Mira	187	M
U Arietis	03	11.0	+14	48	7.2–15.2	Mira	371	M
R Aurigæ	05	17.3	+53	35	6.7–13.9	Mira	458	M
Epsilon Aurigæ	05	02.0	+43	49	2.9– 3.8	Algol	9892	F+B
R Boötis	14	37.2	+26	44	6.2–13.1	Mira	223	M
W Boötis	14	43.4	+26	32	4.7– 5.4	Semi-regular?	450?	M
X Camelopardalis	04	45.7	+75	06	7.4–14.2	Mira	144	K–M
R Cancri	08	16.6	+11	44	6.1–11.8	Mira	362	M
X Cancri	08	55.4	+17	14	5.6– 7.5	Semi-regular	195?	C
R Canis Majoris	07	19.5	−16	24	5.7– 6.3	Algol	1.1	F
S Canis Minoris	07	32.7	+08	19	6.6–13.2	Mira	333	M
R Canum Ven.	13	49.0	+39	33	6.5–12.0	Mira	329	M
R Carinæ	09	32.2	−62	47	3.9–10.5	Mira	309	M
S Carinæ	10	09.4	−61	33	4.5– 9.9	Mira	149	K–M
l Carinæ	09	45.2	−62	30	3.3– 4.2	Cepheid	35.5	F–K
Eta Carinæ	10	45.1	−59	41	−0.8– 7.9	Irregular	–	Pec
R Cassiopeiæ	23	58.4	+51	24	4.7–13.5	Mira	430	M
S Cassiopeiæ	01	19.7	+72	37	7.9–16.1	Mira	612	S
W Cassiopeiæ	00	54.9	+58	34	7.8–12.5	Mira	406	C
Gamma Cass.	00	56.7	+60	43	1.6– 3.0	Irregular	–	B
Rho Cassiopeiæ	23	54.4	+57	30	4.1– 6.2	Semi-regular	–	F–K
R Centauri	14	16.6	−59	55	5.3–11.8	Mira	546	M
S Centauri	12	24.6	−49	26	7–8	Semi-regular	65	C
T Centauri	13	41.8	−33	36	5.5– 9.0	Semi-regular	90	K–M
S Cephei	21	35.2	+78	37	7.4–12.9	Mira	487	C

Star	R.A. h	m	Declination deg.	min.	Range	Type	Period days	Spectrum
T Cephei	21	09.5	+68	29	5.2–11.3	Mira	388	M
Delta Cephei	22	29.2	+58	25	3.5– 4.4	Cepheid	5.4	F–G
Mu Cephei	21	43.5	+58	47	3.4– 5.1	Semi-regular	730	M
U Ceti	02	33.7	−13	09	6.8–13.4	Mira	235	M
W Ceti	00	02.1	−14	41	7.1–14.8	Mira	351	S
Omicron Ceti	02	19.3	−02	59	2.0–10.1	Mira	332	M
R Chamæleontis	08	21.8	−76	21	7.5–14.2	Mira	335	M
T Columbæ	05	19.3	−33	42	6.6–12.7	Mira	226	M
R Comæ Ber.	12	04.3	+18	47	7.1–14.6	Mira	363	M
R Coronæ Bor.	15	48.6	+28	09	5.7–14.8	Irregular	–	C
S Coronæ Bor.	15	21.4	+31	22	5.8–14.1	Mira	360	M
T Coronæ Bor.	15	59.6	+25	55	2.0–10.8	Recurr. nova	–	M+Pec
V Coronæ Bor.	15	49.5	+39	34	6.9–12.6	Mira	358	C
W Coronæ Bor.	16	15.4	+37	48	7.8–14.3	Mira	238	M
R Corvi	12	19.6	−19	15	6.7–14.4	Mira	317	M
R Crucis	12	23.6	−61	38	6.4– 7.2	Cepheid	5.8	F–G
R Cygni	19	36.8	+50	12	6.1–14.4	Mira	426	S
U Cygni	20	19.6	+47	54	5.9–12.1	Mira	463	C
W Cygni	21	36.0	+45	22	5.0– 7.6	Semi-regular	131	M
RT Cygni	19	43.6	+48	47	6.0–13.1	Mira	190	M
SS Cygni	21	42.7	+43	35	7.7–12.4	Dwarf nova	50±	K+Pec
CH Cygni	19	24.5	+50	14	5.6– 9.0	Symbiotic	–	M+B
Chi Cygni	19	50.6	+32	55	3.3–14.2	Mira	408	S
R Delphini	20	14.9	+09	05	7.6–13.8	Mira	285	M
U Delphini	20	45.5	+18	05	5.6– 7.5	Semi-regular	110?	M
EU Delphini	20	37.9	+18	16	5.8– 6.9	Semi-regular	60	M
Beta Doradus	05	33.6	−62	29	3.5– 4.1	Cepheid	9.8	F–G
R Draconis	16	32.7	+66	45	6.7–13.2	Mira	246	M
T Eridani	03	55.2	−24	02	7.2–13.2	Mira	252	M
R Fornacis	02	29.3	−26	06	7.5–13.0	Mira	389	C
R Geminorum	07	07.4	+22	42	6.0–14.0	Mira	370	S
U Geminorum	07	55.1	+22	00	8.2–14.9	Dwarf nova	105±	Pec+M
Zeta Geminorum	07	04.1	+20	34	3.6– 4.2	Cepheid	10.2	F–G
Eta Geminorum	06	14.9	+22	30	3.2– 3.9	Semi-regular	233	M
S Gruis	22	26.1	−48	26	6.0–15.0	Mira	402	M
S Herculis	16	51.9	+14	56	6.4–13.8	Mira	307	M
U Herculis	16	25.8	+18	54	6.4–13.4	Mira	406	M
Alpha Herculis	17	14.6	+14	23	2.7– 4.0	Semi-regular	–	M
68, u Herculis	17	17.3	+33	06	4.7– 5.4	Algol	2.1	B+B
R Horologii	02	53.9	−49	53	4.7–14.3	Mira	408	M
U Horologii	03	52.8	−45	50	6–14	Mira	348	M
R Hydræ	13	29.7	−23	17	3.5–10.9	Mira	389	M
U Hydræ	10	37.6	−13	23	4.3– 6.5	Semi-regular	450?	C
VW Hydri	04	09.1	−71	18	8.4–14.4	Dwarf nova	27±	Pec
R Leonis	09	47.6	+11	26	4.4–11.3	Mira	310	M
R Leonis Minoris	09	45.6	+34	31	6.3–13.2	Mira	372	M
R Leporis	04	59.6	−14	48	5.5–11.7	Mira	427	C
Y Libræ	15	11.7	−06	01	7.6–14.7	Mira	276	M
RS Libræ	15	24.3	−22	55	7.0–13.0	Mira	218	M
Delta Libræ	15	01.0	−08	31	4.9– 5.9	Algol	2.3	A
R Lyncis	07	01.3	+55	20	7.2–14.3	Mira	379	S
R Lyræ	18	55.3	+43	57	3.9– 5.0	Semi-regular	46?	M
RR Lyræ	19	25.5	+42	47	7.1– 8.1	RR Lyræ	0.6	A–F
Beta Lyræ	18	50.1	+33	22	3.3– 4.4	Eclipsing	12.9	B
U Microscopii	20	29.2	−40	25	7.0–14.4	Mira	334	M
U Monocerotis	07	30.8	−09	47	5.9– 7.8	RV Tauri	91	F–K
V Monocerotis	06	22.7	−02	12	6.0–13.9	Mira	340	M
T Normæ	15	44.1	−54	59	6.2–13.6	Mira	241	M
R Octantis	05	26.1	−86	23	6.3–13.2	Mira	405	M
S Octantis	18	08.7	−86	48	7.2–14.0	Mira	259	M
V Ophiuchi	16	26.7	−12	26	7.3–11.6	Mira	297	C

Star	R.A.		Declination		Range	Type	Period	Spectrum
	h	m	deg.	min.			days	
X Ophiuchi	18	38.3	+08	50	5.9– 9.2	Mira	329	M
RS Ophiuchi	17	50.2	−06	43	4.3–12.5	Recurr. nova	–	OB+M
U Orionis	05	55.8	+20	10	4.8–13.0	Mira	368	M
W Orionis	05	05.4	+01	11	5.9– 7.7	Semi-regular	212	C
Alpha Orionis	05	55.2	+07	24	0.0– 1.3	Semi-regular	2335	M
S Pavonis	19	55.2	−59	12	6.6–10.4	Semi-regular	381	M
Kappa Pavonis	18	56.9	−67	14	3.9– 4.8	Cepheid	9.1	G
R Pegasi	23	06.8	+10	33	6.9–13.8	Mira	378	M
Beta Pegasi	23	03.8	+28	05	2.3– 2.7	Irregular	–	M
X Persei	03	55.4	+31	03	6.0– 7.0	Gamma Cass.	–	09.5
Beta Persei	03	08.2	+40	57	2.1– 3.4	Algol	2.9	B
Rho Persei	03	05.2	+38	50	3.3– 4.0	Semi-regular	50?	M
Zeta Phœnicis	01	08.4	−55	15	3.9– 4.4	Algol	1.7	B+B
R Pictoris	04	46.2	−49	15	6.4–10.1	Semi-regular	171	M
L² Puppis	07	13.5	−44	39	2.6– 6.2	Semi-regular	141	M
T Pyxidis	09	04.7	−32	23	6.5–15.3	Recurr. nova	7000±	Pec
U Sagittæ	19	18.8	+19	37	6.5– 9.3	Algol	3.4	B+G
WZ Sagittæ	20	07.6	+17	42	7.0–15.5	Dwarf nova	11900±	A
R Sagittarii	19	16.7	−19	18	6.7–12.8	Mira	270	M
RR Sagittarii	19	55.9	−29	11	5.4–14.0	Mira	336	M
RT Sagittarii	20	17.7	−39	07	6.0–14.1	Mira	306	M
RU Sagittarii	19	58.7	−41	51	6.0–13.8	Mira	240	M
RY Sagittarii	19	16.5	−33	31	5.8–14.0	R Coronæ Bor.	–	G
RR Scorpii	16	56.6	−30	35	5.0–12.4	Mira	281	M
RS Scorpii	16	55.6	−45	06	6.2–13.0	Mira	320	M
RT Scorpii	17	03.5	−36	55	7.0–15.2	Mira	449	S
S Sculptoris	00	15.4	−32	03	5.5–13.6	Mira	363	M
R Scuti	18	47.5	−05	42	4.2– 8.6	RV Tauri	146	G–K
R Serpentis	15	50.7	+15	08	5.2–14.4	Mira	356	M
S Serpentis	15	21.7	+14	19	7.0–14.1	Mira	372	M
T Tauri	04	22.0	+19	32	9.3–13.5	Irregular	–	F–K
SU Tauri	05	49.1	+19	04	9.1–16.9	R Coronæ Bor.	–	G
Lambda Tauri	04	00.7	+12	29	3.4– 3.9	Algol	4.0	B+A
R Trianguli	02	37.0	+34	16	5.4–12.6	Mira	267	M
R Ursæ Majoris	10	44.6	+68	47	6.5–13.7	Mira	302	M
T Ursæ Majoris	12	36.4	+59	29	6.6–13.5	Mira	257	M
U Ursæ Minoris	14	17.3	+66	48	7.1–13.0	Mira	331	M
R Virginis	12	38.5	+06	59	6.1–12.1	Mira	146	M
S Virginis	13	33.0	−07	12	6.3–13.2	Mira	375	M
SS Virginis	12	25.3	+00	48	6.0– 9.6	Semi-regular	364	C
R Vulpeculæ	21	04.4	+23	49	7.0–14.3	Mira	137	M
Z Vulpeculæ	19	21.7	+25	34	7.3– 8.9	Algol	2.5	B+A

Mira Stars: maxima, 1992

JOHN ISLES

Below are given predicted dates of maxima for Mira stars that reach magnitude 7.5 or brighter at an average maximum. Individual maxima can in some cases be brighter or fainter than average by a magnitude or more, and all dates are only approximate. The positions, extreme ranges and mean periods of these stars can all be found in the preceding list of interesting variable stars.

Star	Mean magnitude at max.	Dates of maxima
R Andromedæ	6.9	Oct. 16
W Andromedæ	7.4	Jan. 11
R Aquilæ	6.1	July 16
R Boötis	7.2	May 14, Dec. 24
R Cancri	6.8	Nov. 2
S Canis Minoris	7.5	Sep. 6
R Carinæ	4.6	Apr. 25
S Carinæ	5.7	Mar. 27, Aug. 23
R Cassiopeiæ	7.0	June 23
T Cephei	6.0	Oct. 15
U Ceti	7.5	Mar. 26, Nov. 16
Omicron Ceti	3.4	July 14
T Columbæ	7.5	Apr. 1, Nov 13
S Coronæ Borealis	7.3	Nov. 3
V Coronæ Borealis	7.5	June 1
R Corvi	7.5	June 26
R Cygni	7.5	Oct. 14
U Cygni	7.2	Mar. 31
RT Cygni	7.3	May 30, Dec. 7
Chi Cygni	5.2	Mar. 22
R Geminorum	7.1	Oct. 16
U Herculis	7.5	Feb. 1
R Horologii	6.0	Mar. 21
U Horologii	7	Mar. 23
R Leonis	5.8	Aug. 31
R Leonis Minoris	7.1	July 20
RS Libræ	7.5	Feb. 23, Sep. 27
V Monocerotis	7.0	Jan. 23, Dec. 29
T Normæ	7.4	May 9
V Ophiuchi	7.5	Jan. 26, Nov. 18
X Ophiuchi	6.8	Apr. 6
U Orionis	6.3	Dec. 1

R Sagittarii	7.3	Apr. 20
RR Sagittarii	6.8	May 18
RT Sagittarii	7.0	July 26
RU Sagittarii	7.2	Jan. 20, Sep. 17
RR Scorpii	5.9	June 17
RS Scorpii	7.0	Oct. 26
S Sculptoris	6.7	Oct. 6
R Serpentis	6.9	Apr. 5
R Trianguli	6.2	Mar. 22, Dec. 14
R Ursæ Majoris	7.5	Oct. 4
R Virginis	6.9	Jan. 9, June 3, Oct. 27
S Virginis	7.0	Aug. 6

Some Interesting Double Stars

R. W. ARGYLE

Name	Magnitudes	Separation in seconds of arc	Position angle, degrees	Remarks
Gamma Andromedæ	2.3, 5.0	9.4	064	Yellow, blue. B is again double
Zeta Aquarii	4.3, 4.5	1.9	217	Slowly widening.
Gamma Arietis	4.8, 4.8	7.8	000	Very easy. Both white.
Theta Aurigæ	2.6, 7.1	3.5	313	Stiff test for 3-in.
Delta Boötis	3.5, 8.7	105	079	Fixed.
Epsilon Boötis	2.5, 4.9	2.8	335	Yellow, blue. Fine pair.
Zeta Cancri	5.6, 6.2	5.8	085	A again double.
Iota Cancri	4.2, 6.6	31	307	Easy. Yellow, blue.
Alpha Canum Ven.	2.9, 5.5	19.6	228	Easy. Yellow, bluish.
Alpha Capricorni	3.6, 4.2	379	291	Naked-eye pair.
Eta Cassiopeiæ	3.4, 7.5	12.5	312	Easy. Creamy, bluish.
Beta Cephei	3.2, 7.9	14	250	Easy with a 3-in.
Delta Cephei	var, 7.5	41	192	Very easy.
Alpha Centauri	0.0, 1.2	19.7	214	Very easy. Period 80 years.
Xi Cephei	4.4, 6.5	6.3	270	White, blue.
Gamma Ceti	3.5, 7.3	2.9	294	Not too easy.
Alpha Circini	3.2, 8.6	15.7	230	PA slowly decreasing.
Zeta Corona Bor.	5.1, 6.0	6.3	305	PA slowly increasing.
Delta Corvi	3.0, 9.2	24	214	Easy with a 3-in.
Alpha Crucis	1.4, 1.9	4.2	114	Third star in a low power field.
Gamma Crucis	1.6, 6.7	124	024	Third star in a low power field.
Beta Cygni	3.1, 5.1	34.5	055	Glorious. Yellow, blue.
61 Cygni	5.2, 6.0	30	148	Nearby binary. Period 722 years.
Gamma Delphini	4.5, 5.5	9.6	268	Easy. Yellowish, greenish.
Nu Draconis	4.9, 4.9	62	312	Naked eye pair.
Alpha Geminorum	1.9, 2.9	3.0	77	Widening. Visible with a 3-in.
Delta Geminorum	3.5, 8.2	6.5	120	Not too easy.
Alpha Herculis	var, 5.4	4.6	106	Red, green.
Delta Herculis	3.1, 8.2	9.8	272	Optical pair. Distance increasing.
Zeta Herculis	2.9, 5.5	1.6	084	Fine, rapid binary. Period 34 years.
Gamma Leonis	2.2, 3.5	4.4	123	Binary, 619 years.
Alpha Lyræ	0.0, 9.5	71	180	Optical pair. B is faint.

Name	Magnitudes	Separation in seconds of arc	Position angle, degrees	Remarks
Epsilon[1] Lyræ	5.0, 6.1	2.6	356	Quadruple system. Both
Epsilon[2] Lyræ	5.2, 5.5	2.2	093	pairs visible in a 3-in.
Zeta Lyræ	4.3, 5.9	44	149	Fixed. Easy double.
70 Ophiuchi	4.2, 6.0	1.5	224	Rapid motion.
Beta Orionis	0.1, 6.8	9.5	202	Can be split with a 3-in.
Iota Orionis	2.8, 6.9	11.8	141	Enmeshed in nebulosity.
Theta Orionis	6.7, 7.9	8.7	032	Trapezium in M42.
	5.1, 6.7	13.4	061	
Sigma Orionis	4.0, 10.3	11.4	238	Quintuple. A is a
	6.5, 7.5	30.1	231	close double.
Zeta Orionis	1.9, 4.0	2.4	162	Can be split in 3-in.
Eta Persei	3.8, 8.5	28.5	300	Yellow, bluish.
Beta Phœnicis	4.0, 4.2	1.5	324	Slowly widening.
Beta Piscis Aust.	4.4, 7.9	30.4	172	Optical pair. Fixed.
Alpha Piscium	4.2, 5.1	1.9	278	Binary, 933 years.
Kappa Puppis	4.5, 4.7	9.8	318	Both white.
Alpha Scorpii	1.2, 5.4	3.0	275	Red, green.
Nu Scorpii	4.3, 6.4	42	336	Both again double.
Theta Serpentis	4.5, 5.4	22.3	103	Fixed. Very easy.
Alpha Tauri	0.9, 11.1	131	032	Wide, but B very faint in small telescopes.
Beta Tucanæ	4.4, 4.8	27.1	170	Both again double.
Zeta Ursæ Majoris	2.3, 4.0	14.4	151	Very easy. Naked eye pair with Alcor.
Xi Ursæ Majoris	4.3, 4.8	1.3	060	Binary, 60 years. Closing. Needs a 4-in.
Gamma Virginis	3.5, 3.5	3.0	287	Binary, 171 years. Closing.
Theta Virginis	4.4, 9.4	7.1	343	Not too easy.
Gamma Volantis	3.9, 5.8	13.8	299	Very slow binary.

Some Interesting Nebulæ and Clusters

Object	R.A.		Dec.		Remarks
	h	m			
M.31 Andromedæ	00	40.7	+41	05	Great Galaxy, visible to naked eye.
H.VIII 78 Cassiopeiæ	00	41.3	+61	36	Fine cluster, between Gamma and Kappa Cassiopeiæ.
M.33 Trianguli	01	31.8	+30	28	Spiral. Difficult with small apertures.
H.VI 33–4 Persei	02	18.3	+56	59	Double cluster; Sword-handle.
△142 Doradûs	05	39.1	−69	09	Looped nebula round 30 Doradûs. Naked-eye. In Large Cloud of Magellan.
M.1 Tauri	05	32.3	+22	00	Crab Nebula, near Zeta Tauri.
M.42 Orionis	05	33.4	−05	24	Great Nebula. Contains the famous Trapezium, Theta Orionis.
M.35 Geminorum	06	06.5	+24	21	Open cluster near Eta Geminorum.
H.VII 2 Monocerotis	06	30.7	+04	53	Open cluster, just visible to naked eye.
M.41 Canis Majoris	06	45.5	−20	42	Open cluster, just visible to naked eye.
M.47 Puppis	07	34.3	−14	22	Mag. 5,2. Loose cluster.
H.IV 64 Puppis	07	39.6	−18	05	Bright planetary in rich neighbourhood.
M.46 Puppis	07	39.5	−14	42	Open cluster.
M.44 Cancri	08	38	+20	07	Præsepe. Open cluster near Delta Cancri. Visible to naked eye.
M.97 Ursæ Majoris	11	12.6	+55	13	Owl Nebula, diameter 3'. Planetary.
Kappa Crucis	12	50.7	−60	05	'Jewel Box'; open cluster, with stars of contrasting colours.
M.3 Can. Ven.	13	40.6	+28	34	Bright globular.
Omega Centauri	13	23.7	−47	03	Finest of all globulars. Easy with naked eye.
M.80 Scorpii	16	14.9	−22	53	Globular, between Antares and Beta Scorpionis.
M.4 Scorpii	16	21.5	−26	26	Open cluster close to Antares.
M.13 Herculis	16	40	+36	31	Globular. Just visible to naked eye.
M.92 Herculis	16	16.1	+43	11	Globular. Between Iota and Eta Herculis.
M.6 Scorpii	17	36.8	−32	11	Open cluster; naked eye.
M.7 Scorpii	17	50.6	−34	48	Very bright open cluster; naked eye.
M.23 Sagittarii	17	54.8	−19	01	Open cluster nearly 50' in diameter.
H.IV 37 Draconis	17	58.6	+66	38	Bright Planetary.
M.8 Sagittarii	18	01.4	−24	23	Lagoon Nebula. Gaseous. Just visible with naked eye.
NGC 6572 Ophiuchi	18	10.9	+06	50	Bright planetary, between Beta Ophiuchi and Zeta Aquilæ.
M.17 Sagittarii	18	18.8	−16	12	Omega Nebula. Gaseous. Large and bright.
M.11 Scuti	18	49.0	−06	19	Wild Duck. Bright open cluster.
M.57 Lyræ	18	52.6	+32	59	Ring Nebula. Brightest of planetaries.
M.27 Vulpeculæ	19	58.1	+22	37	Dumb-bell Nebula, near Gamma Sagittæ.
H.IV 1 Aquarii	21	02.1	−11	31	Bright planetary near Nu Aquarii.
M.15 Pegasi	21	28.3	+12	01	Bright globular, near Epsilon Pegasi.
M.39 Cygni	21	31.0	+48	17	Open cluster between Deneb and Alpha Lacertæ. Well seen with low powers.

Our Contributors

Dr David Allen is well known to all *Yearbook* readers. He continues his work at the Anglo-Australian Observatory in New South Wales, though this year's article from him was compiled as a result of a visit to Arizona!

Dr Allan Chapman, of the Centre for Mediæval and Renaissance Studies, Oxford, is very well known as a scientific historian, and has published many research papers.

John Fletcher is an amateur astronomer who lives at Tuffley in Gloucestershire, where he has his private observatory. His main interests are deep-sky photography and the search for supernovæ, as well as observations of Mars and Venus.

Dr David Gavine is a leading observer of auroræ, and is well known for his researches. He lives in Edinburgh.

Ron Ham is an amateur radio astronomer who lives in Sussex, where he has established his solar observatory. He has written many research papers.

John Isles is a leading amateur observer of variable stars. For many years he has been Director of the Variable Star Section of the British Astronomical Association, and has also acted as the Association's Vice-president. He now lives in Cyprus.

Dr R. C. Maddison has been for many years in charge of the Astronomy Department at Keele University. He has now left to live in Merritt Island, Cape Canaveral (Florida) and is associated with the Astronauts Memorial Science Center in Cocoa.

Dr Paul Murdin, one of our most regular and valued contributors recently left the Royal Greenwich Observatory, where he had been Deputy Director, to assume the Directorship of the Royal Observatory, Edinburgh, in succession to Professor Malcolm Longair.

The William Herschel Society maintains the museum now established at 19 New King Street, Bath – the only surviving Herschel House. It also undertakes activities of various kinds. New members would be welcome; those interested are asked to contact Dr L. Hilliard at 2 Lambridge, London Road, Bath.